PENGUIN

DISCOURSE
AND THE
MEDITATIONS

RENÉ DESCARTES was born in 1596 at La Haye near Tours, and educated at the Jesuit College at La Flèche. Like many of his generation he contested the value of an education based on Aristotelianism and, after leaving college, he attempted to resolve the sceptical crisis of his age by inventing a method of reasoning based on mathematics. After serving as a soldier in Holland, Bohemia and Hungary, he left the army in 1621 and devoted himself to science and philosophy. In 1629 he retired to Holland where he lived and worked in great seclusion for twenty years. However, his doctrines involved him in some bitter arguments with Dutch theologians, and in 1648 he accepted an invitation from Queen Christina of Sweden to instruct her in philosophy. He died in Stockholm in 1650.

·

F. E. SUTCLIFFE, Chevalier de l'Ordre National du Mérite, was Professor of Classical French Literature at the University of Manchester from 1966 until 1982. He joined the staff of that University in 1946, after serving for six years in the Royal Artillery. He published *La Pensée de Paul Valéry* (1954), *Guez de Balzac et son temps, Littérature et politique* (1959), *Le Réalisme de Charles Sorel, Problèmes humains du XVIIe siècle* (1965), an edition of the *Discours politiques et militaires*, of François de la Noue (1967) and *Politique et culture 1560–1660* (1973). Professor Sutcliffe died in 1983.

Descartes

DISCOURSE ON METHOD
AND THE
MEDITATIONS

*

Translated with an Introduction
by F. E. Sutcliffe

PENGUIN BOOKS

PENGUIN BOOKS

Published by the Penguin Group
27 Wrights Lane, London w8 5tz, England
Viking Penguin Inc., 40 West 23rd Street, New York, New York 10010, USA
Penguin Books Australia Ltd, Ringwood, Victoria, Australia
Penguin Books Canada Ltd, 2801 John Street, Markham, Ontario, Canada l3r 1b4
Penguin Books (NZ) Ltd, 182–190 Wairau Road, Auckland 10, New Zealand

Penguin Books Ltd, Registered Offices: Harmondsworth, Middlesex, England

This translation first published 1968
21 23 25 27 29 30 28 26 24 22

Made and printed in Great Britain by
Richard Clay Ltd, Bungay, Suffolk
Set in Monotype Garamond

CONTENTS

INTRODUCTION

BORN on 31 March 1596 at La Haye in Touraine, Descartes was the third son of a country gentleman who, after seeing service in the army, had become councillor of the *Parlement* of Britanny. At the age of ten he entered the Jesuit college of La Flèche where he stayed for eight years. The University, which had been so brilliant in the sixteenth century, had fallen to a low level as a consequence of the wars of religion and the successive purges which had led to an atmosphere of suspicion not conducive to creative activity. The Jesuits put their hands on the colleges with remarkable speed and within a very short time had a virtual monopoly of the education of the ruling classes. The Jesuit order was young and essentially modern, not limiting itself to theology, but devoting part of its time to literary studies and to profane sciences. In borrowing thus the arms of humanism, it set out to be a militant order with a high degree of culture whose members could mingle in society and discuss on a footing of equality with the lay intelligentsia.

In its colleges the order laid the greatest emphasis on pedagogical training, unlike the university of the Renaissance where erudition had had pride of place. In other words the idea of *method* was at the centre of its educational practice. Everything was done according to rules: the way of placing one's feet, of deporting oneself, of speaking, and so on; but it is important to remember that this preoccupation with method did not denote a spirit of routine and idleness. On the contrary, it arose from a desire to achieve the efficacious, from a will to succeed. In spite of his criticism of an education of which the basis was constituted by the classical humanities (the sciences,

7

although not excluded, were envisaged from a purely practical angle and were directed towards training in the art of fortification), Descartes had nothing but respect for the actual methods of the Jesuits and throughout his life he sought their approval of his work. It is, moreover, indicative of the influence exercised upon him by these methods that his first writing, which we no longer have, was a treatise on the rules of fencing and his second an examination of the mathematical basis of harmony.

In 1617, like many compatriots of his class, he went to learn the art of war in the army of Prince Maurice of Nassau at Breda. After two years of service, disappointed by the lack of opportunities to fight, he moved to Germany and to the army of the Duke of Bavaria, Maximilian, but here too inactivity was all that he found and he finally left the army in 1621. These, however, were determining years for his philosophical development. It was Beeckman, the Dutch mathematician, who, impressed by the vigour of the young Descartes' intelligence, advised him to concentrate his attention on the problems of mathematical physics. Descartes seized upon the idea with enthusiasm and, as early as 1619, was writing to Beeckman: 'What I wish to finish is ... an absolutely new science enabling one to resolve all questions proposed on any order of continuous or discontinuous quantities.' The whole of Cartesian philosophy is contained in embryo in this phrase. However, he continued, during the succeeding nine years, to disperse his efforts, interesting himself in medicine, chemistry and optics, and to travel widely. It was only after his conversation with the founder of the Oratory, Cardinal de Bérulle in 1627, that he finally set to work. During a meeting of philosophers and theologians, Descartes had spoken of the ideas which he had nurtured in 1619. Bérulle encouraged him to pursue his meditations, to use them to serve the faith and to communicate them to others. Shortly after this, Descartes decided to settle in Holland, away from the distractions of Parisian society, and to devote himself, undisturbed, to the task indicated to him by Bérulle. The first exposition of his philosophy, *Rules for the Direction of the Mind*, published only after his death, was composed in 1628. Then, with frequent interruptions, hesitations and dispersions, for the next five years he worked on what was to become a veritable scientific *summa*, entitled *De Mundo*, and described in Chapter 5 of the *Discourse*, a *summa* which in his mind was to serve as an infrastructure to modern Christian thought, in the same way as the philosophy of Aristotle had served medieval Christian thought through the synthesis of St Thomas Aquinas. The *De Mundo* was ready for publication when, in November 1633, Descartes heard belatedly of the condemnation of Galileo. Now at the basis of Descartes' system was the Copernican theory of the rotation of the earth: to publish, in face of the attitude of the Church, would be to incur, not any real danger in a relatively liberal France, but the risk of failure in the pursuit of the aim which he had set himself, namely to see his philosophy accepted and taught by those best placed to disseminate it, the Jesuits. His disappointment was intense. Six months later, however, he wrote to his friend Mersenne that all hope was not lost, that what then appeared heretical would one day cease to do so, and that his *De Mundo* would, in time, be publishable. It was then that he conceived the project of testing, as it were, the defences of the enemy by publishing a few samples of the *De Mundo*. At the same time he hoped that by addressing himself to the generality of the educated but not specialized public, he would create a favourable current of opinion and find himself solicited to reveal more of his work. He first decided to publish the *Dioptric*, the chapter in which he studies the nature of light and of refraction;

he next took the decision to add to this the treatise on
Meteors, in which he exposes his theory of matter, and to
present these two with a preface to the public. Finally, he
added the essay in which he explains his new analytic
geometry. The whole work preceded by a preface – the
Discourse on Method – appeared in 1637. Four years later
he gave in Latin a full exposition of his metaphysics, a
brief account of which had been given in Chapter 4 of the
Discourse. This time he was addressing himself to the
theologians in their own language: the work, entitled
Meditationes de Prima Philosophia, was dedicated to the
Dean and Doctors of the Sacred Faculty of Theology of
Paris and was intended to demonstrate that Descartes'
philosophy was superior to the scholastic in so far that it
not only dispensed with the probabilities and verisimilitudes of scholasticism, replacing them with arguments
which presented the same degree of certainty as a mathematical proof, but also led to perfectly orthodox conclusions. The *Meditations* gave rise to a multiplicity of attacks:
Arnauld, Gassendi and Hobbes in particular raised objections and to these Descartes replied in the second edition
of his work. His last two publications were the *Principles
of Philosophy* and *The Passions of the Soul*. The *Principles*,
published in 1644, restated in the first part the metaphysics
of the *Meditations*, while the remainder, addressed to the
scientific world, contained a description of the structure
of the universe and an account of the relation between
body and soul. Three years later the *Principles* were translated into French and published with a preface in the form
of a letter addressed to the translator in which Descartes,
with bold assurance, states what he considers to be the
true role of philosophy, and recalls the extent of his own
success in the application of his system. His metaphysical
and scientific work was completed. The time which remained to him was directed to an attempt to apply his

method to medicine and ethics. That human happiness is conditioned by the progress of medicine, Descartes never doubted, and he never ceased to preoccupy himself with the problem. But the constitution of medicine as a deductive science revealed itself as more difficult than that of ethics and it was to this subject in particular that he turned. He corresponded on the subject with Princess Elisabeth of Bohemia, and composed for Queen Christine of Sweden his *Passions of the Soul* which contains an exposition of his ethics conceived as science.

In 1649 he accepted an invitation from the Queen to go to the Swedish Court in Stockholm to instruct her in his philosophy. The unaccustomed cold and the necessity, imposed by the queen, of giving her lessons at five in the morning proved too much for the philosopher's health. He contracted inflammation of the lungs and died on 11 February 1650.

*

The *Discourse on Method*, one of the most famous texts in the French language, presents something of a paradox. The work of a thinker, who whatever else is known of him, has for three centuries been considered as the prototype of clarity, it is curiously obscure, its plan a caricature of logical composition. The first section of this preface to three scientific treatises contains a biography; the second a methodological exposition which, instead of being continued by details of its application to the sciences, is followed by a chapter on ethics and another on metaphysics. The thread is renewed in the fifth chapter, after which the sixth and last forms a sort of new introduction taking up themes already treated.

Gadoffre* has established beyond all doubt the various

---

* Descartes: *Discours de la Méthode, avec introduction et remarques de Gilbert Gadoffre*, Manchester University Press, 1941.

stages of the composition of this text. According to him the sixth chapter constitutes the original preface, being composed before the *Geometry* of which it contains no mention. A further proof of this assertion is that the methodology of Chapter 2 is not mentioned and the stoic maxims of the third chapter are frankly contradicted. As for Chapter 1, it is almost certainly a final draft of a work which, according to Guez de Balzac, Descartes had sent to him in 1628, and entitled a *History of My Mind*. Chapter 2 was composed after the *Geometry* which it illuminates. Finally, Chapter 3 was composed last of all as an afterthought. The Chancellor, Séguier, having refused to grant the privilege necessary for publication until the full text of the *Discourse* was submitted to him, Descartes, in order to forestall any possible objections, hastily added the third chapter, a manifesto of political orthodoxy and an antidote to the revolutionary virulence which some might have discerned in the maxim according to which one should, once in one's life, rid oneself of all the opinions which one had hitherto harboured in one's mind. Not indeed that Descartes was not sincerely conservative in matters of politics. He most certainly was, in common with the majority of the most advanced thinkers of his time, but the circumstance of the composition of this chapter must of necessity lead one to hesitate to take any account of it in a study of Descartes' thinking on ethics.

The *Discourse* then presents a heterogeneous character, for various historical and psychological reasons, and yet, there is another and not less important reason for this phenomenon. Descartes had a very clear idea of the type of audience which he was trying to reach: that of the cultured public of society, the ladies of the 'salons' rather than the pedants of the University. In other words, the public which had made the extraordinary success of his friend Guez de Balzac, whose letters and essays had brought within the reach of the public of the salons matters which, up to then, had been confined to the attention of specialists and whose art consisted in dealing with serious subjects in an agreeable and worldly tone far removed from that of strictly erudite circles. Balzac spoke, when treating the question of literary practice, of the necessity of 'gilding and perfuming', of winning over the minds of his readers by appealing to their senses. Descartes, who had great admiration for Balzac, consciously set out to imitate his example.

Further, by addressing his reader as one individual to another, Descartes was accomplishing three important functions. First he was satisfying one of the principle interests of man in the post-medieval period: interest in human psychology, in the relations which exist between the individual and the world outside him. The *Essays* of Montaigne, the innumerable volumes of *Mémoires* of soldiers and diplomats, the descriptions of the inhabitants of foreign lands written both by missionaries and political theorists, the emphasis placed on the study of history as being an introduction to the understanding of practical living, all bear witness to a passionate interest in the individual and in his relations with his surroundings. Descartes could be assured of striking a deep note of sympathy in his reader by his account of his reactions to his school-days: they were indeed those of many of his contemporaries dissatisfied no less than he with the promises of masters attached to the out-dated values of the Greco-Latin humanities. Right from the outset of his *Discourse*, therefore, Descartes places his reader, the reader of 1637, in a situation with which he is familiar and confronts him with problems which are also his problems. What could stimulate curiosity more than to know how Descartes had resolved them? Secondly, the ultimate aim of Descartes was to persuade men that, in their task of reconstructing the world, a method, his method, was alone effective. That is to say that his method was essentially an instrument for action. Those civilizations which applied it would progress more quickly than those which did not. Since, therefore, Descartes' thought tends constantly towards that which is useful for life, for individual and collective well-being, it is evident that he should have every reason to present his method as having been created in contact with concrete reality. Born of reflection on living experience, it is intended to authorize a return to experience at a higher level of awareness and ultimately to a complete domination of the real. Lastly, from a purely tactical point of view, this method has the advantage of attenuating the universalizing tendencies contained in the Cartesian system. By emphasizing its personal origins and character, Descartes is taking a necessary precaution. Hence the peculiar character of the treatise, its unique tone: a tone of enthusiasm, of joy in struggle and triumph. Descartes presents his spiritual itinerary as an adventure during which a series of obstacles are encountered, each to be overcome in its turn until total victory is obtained. One is struck by the number of military images which come from his pen and one rightly guesses that we are here in the presence not of a professional philosopher but of a soldier who, with a remarkable audacity and a great nobility of spirit, sets out upon the path of intellectual conquest.

While bringing notoriety to Descartes the *Discourse* did not have the type of success which he had hoped for. Five hundred copies were printed, of which two hundred were reserved to the author. These he distributed among the Jesuits, diplomats and leading members of Parisian society. And yet, at his death in 1650, the three hundred remaining copies were still not exhausted. In 1647 Descartes had to agree to give a Latin version in order to reach the specialized reader. Those of his works which had the greatest success were the *Principles*, addressed to the scientists, and *The Passions of the Soul*, written for Christina of Sweden and not intended for publication.

*

The essential elements of Cartesianism are contained in Chapters 1, 2, 4 and 5.

Chapter 1 contains a vigorous attack on an education based on the humanities and in particular on the official Aristotelian philosophy which, says Descartes, provides one with the means of talking about all and everything in terms of probabilities, of verisimilitudes. A multitude of opinions, each equally probable, are admitted on any given subject, and since the sciences borrow their principles from philosophy, they too present no firm basis of truth. In belief, Descartes is seeking for certainties, and amongst all the matters which he reviews: ethics, history, medicine, philosophy, he finds only two which provide such certainties: theology and mathematics. But the truths of religion are revealed truths, inaccessible to our intelligence; there remain therefore only the truths of mathematics which, he states, have up to now been used only for the mechanical arts, whereas they could serve as the basis of something more noble.

Chapter 2 contains the development of this idea. With infinite precaution Descartes leads his reader towards acceptance of the notion that the unity of the sciences can be achieved, and achieved only by Descartes. A building constructed by one architect, a fortification by one engineer, a legal system by one legislator are more perfect than those constructed by many. Similarly the scholastic philosophy suffers from the glosses of innumerable persons, and human opinions offer no certain truth from the mere fact that they are multiple and diverse: the philosophers disagree with each other, opinions vary from country to country and from time to time, according to the whim of fashion. But he, Descartes, thanks to his study of mathematics, has found a method which, if properly used, is capable of leading to certainty. At the basis of his thought is the notion of the unity of mathematics, and by extension, the unity of all the sciences. By his coordinate geometry, Descartes could demonstrate how geometry and algebra dovetail each other, how an interchange of ideas is possible by the identification of algebraic correlation with geometrical locus. Numerical relationship can be expressed as a spatial one, lines are changed into numbers and numbers into lines. Similarly, the notion of order in mathematical progression is essential. Order exists where the knowledge of the term follows of necessity from knowledge of another. This is not the order of the Scholastics which is merely classificatory – serving, as Descartes says, merely to explain to others what one already knows – but a dynamic order leading to the discovery of the unknown terms. By extrapolating his practice in the field of mathematics, Descartes establishes four rules of method which he presents as valid for the study of all sciences, since what is important is not the objects of the individual sciences, but the operations of the inquiring mind which are everywhere the same.

The four famous rules require some explanation. The first implies the operation which Descartes knows as 'intuition', that is to say the use of the pure light of the mind as opposed to the evidence of the senses or of the imagination. It is by 'intuition' that each man knows that he is, that he thinks, that a triangle has three sides. This first rule therefore may be paraphrased thus: in the study of any problem, start by embracing 'intuitively' the fundamental truths of which there can be no doubt. The second, often known as the rule of analysis, enjoins to decompose complex problems into problems as simple as possible. The third, known as the rule of synthesis, applies to the truths reached by the two preceding rules. Put them in order, says Descartes, starting with the simplest, those reached by application of the first rule, followed by the truths deduced from them, going from the simpler to the more complex. It is obvious that there is here a direct application of the principle of the formation of equations, of the movement from equations of the first degree to those of a higher degree. Lastly, the fourth rule takes account of the fact that deduction, unlike intuition, depends to some extent on memory. In order to guarantee oneself against any defect of memory, one should, says Descartes in this rule, attempt to give to deduction the character of intuition by exercising oneself to see immediately, in a deductive reasoning, the link between the first principles and their ultimate consequences.

This method will allow one to operate with the same success when dealing with the external world as it does when dealing with mathematical objects, for mathematical relations are of the same order as those of the understanding, and the external world is also mathematical in its structure. Now the hypothesis of the mathematism of nature is by no means peculiar to Descartes; the leading scientists of his day were no less than he convinced of its validity. But, for Descartes, this is merely the premise of a deductive science: from the notion of space and movement, he wishes to reconstruct the whole universe. His physics being consequently a physics of ideas, he needs to give to it a metaphysical basis in order to acquire the assurance that to the idea there corresponds an external reality, hence the important fourth chapter.

Here he enters upon the second phase of his quest for truth. He has found clear and distinct ideas, he has the clear and distinct idea of a triangle, for example, but he has as yet no guarantee of the real existence of the triangle. Aristotle had started from the given, outside world, in all its complexity and full of all the qualities which sense perception discerns in it. Descartes, on the other hand, had rejected the validity of the evidence of the senses: we cannot say of a material object that it has the property of heat or of cold, for these are not clear and distinct ideas. The only clear idea one can have of objects is that they are extended in height, depth and breadth, that is to say the idea of them which can be expressed mathematically. Matter is identified with geometrical extension. So far, therefore, from proceeding as did Aristotle from complex reality to unifying principles, Descartes proceeds in the opposite direction, from the idea to the thing. But the clarity of an idea does not entail, of necessity, the existence of its object. Descartes has as yet no assurance that the real is not, in fact, irrational and obscure. In asserting the primacy of clear and distinct ideas, in reducing the material world to nothing but extension and movement, he has expressed an act of faith. Is he justified in doing so? To this question Chapter 4 provides an attempt at an answer.

How is he to proceed? By rejecting, he says, as being absolutely false everything of which he should have the slightest cause to doubt, and then to see if there remained anything which was entirely indubitable. He uses the same arguments of the Sceptics as Montaigne had used: criticism of the senses, criticism of reason. Our senses often deceive us; we often make mistakes in reasoning even on the simplest matters of geometry. But whereas Montaigne had concluded that the Sceptics had been right in asserting that the human mind is incapable of reaching any certainty, Descartes, at the moment when all issue appears closed, brings forward dramatically his proposition 'I

think therefore I am' (*cogito ergo sum*). The great originality
of Descartes, and that which enables him to avoid the
conclusion of Montaigne and the Sceptics is that, instead
of considering the objects of doubt, he detaches the act of
doubting from reference to anything external to itself, and
in that way cuts the ground from beneath the feet of
scepticism. For doubting is thinking and is therefore
linked to his existence. He cannot perceive that he thinks
without at the same time being certain that he is. 'I am'
is the inevitable concomitant of 'I think'. Thus it is that
Descartes can formulate a judgement of existence: I exist
as a thinking being.

The Cogito is a first principle from which Descartes
will now deduce all that follows. He has a clear and distinct conception of the fact that he exists; he can therefore
believe that whatever else he perceives with the same
clarity and distinction is equally true. Moreover, he knows
himself only as a thinking being, he is therefore assured
that the soul and the body are entirely distinct. Since he
has been able to understand his own being and essence
without yet knowing anything about the world outside
him, it follows that his self – or soul – is completely
independent of the outside world, mind is distinct from
and superior to matter. Next, by reflecting on the notion
of doubt itself, Descartes perceives that, as to know is a
greater perfection than to doubt, he must be an imperfect
being. But an imperfect being cannot produce the idea of
perfection which he nevertheless possesses. No other
being, imperfect like himself, could have given it to him;
only a perfect being could have done so, therefore a
perfect being, God, exists. Two other proofs of the existence of God follow. Descartes exists, possesses the idea of
perfection, and is himself, imperfect. If he had created
himself, he would have created himself perfect. He has
therefore been created by another who must of necessity

be perfect since Descartes has the idea of perfection. Lastly the famous ontological proof, the most important. Absolute perfection is the sum total of all possible perfections. God, being absolute perfection, must necessarily exist since existence is one of the perfections. To imagine that God does not exist is as absurd as to imagine a triangle which has not three angles. Descartes was very pleased with this proof and wrote to Mersenne: 'I dare to boast that I have found a proof of the existence of God which I find fully satisfactory and by which I know that God exists more certainly than I know the truth of any geometrical proposition.' And indeed this proof has, inside the framework of Cartesianism, a notable advantage, in so far that it is apprehended by the intuition rather than by deductive reasoning, and therefore presents a greater degree of evidence. Once the existence of God has been established, it is easy to show that, as a perfect being, he cannot deceive us and that consequently we can place our confidence in the veracity of our clear and distinct ideas.

Little need be said of Cartesian physics; its deductive character often led Descartes to over-hasty generalizations. Its influence was limited and it was rapidly superseded by the physics of Newton. It is, however, important to note that it marks, at its date, the most complete break with the Aristotelian and medieval conception of the cosmos. Aristotelian physics had been based on sense perception. Nature actually possessed the forces and qualities which we seem to discern in it. Moreover, everything in the cosmos was characterized by a greater or lesser degree of value or perfection according to a hierarchical scheme, going from matter at the foot to God, the first mover and ultimate end, at the summit. At the centre was the earth, and centred upon earth was man, all the other contents of the universe being ordered around

him and for him. By excluding all forms, qualities and
forces, and by reducing matter to its mathematical expression, Descartes, at the same time, ruins the very notion of
the ancient cosmos. Henceforward, the only spectacle
which presents itself to the inquiring eye of man is that of
matter agitated by movements according to mathematical
laws. God is no longer present in the world and neither
is man in the sense that he no longer has an assigned place
there. As mind, infinitely separated from a world which
is matter, the role of man can only be that of dominating
his surroundings, of becoming 'master and possessor of
Nature'. However much the Cartesian dualism of mind
and matter has bedevilled philosophy, it opened the doors
to the development of modern science.

*

The system of Descartes is a reply, not only to the system
of Aristotle, but also, and perhaps even primarily, to the
naturalism of the preceding century, that of Pomponazzi,
of Bruno and Vanini, that of the astrologers and alchem-
ists for whom nature was animated by a soul. Lenoble*
has shown into what error one falls in considering that
the naturalists, simply because they lay emphasis on
experience and deny the medieval notion of the miracu-
lous, are the forerunners of modern scientific thought.
On the contrary, they represent a step backwards in relation to Aristotle in so far that they intensify and generalize
the action of the occult. What characterizes the men of the
generation of Descartes is above all the will to dominate,
to control events, to eliminate chance and the irrational.
This attitude is present in every field: the political, the
military, the scientific. But how can one control phenomena if one cannot foresee the way in which phenomena
will behave? For Machiavelli chance still controlled over

* R. Lenoble: *Mersenne, ou la naissance du mécanisme*, Paris, 1943.

half of events, leaving us the control of the remainder. The elimination of chance becomes an indispensable condition of man's supremacy. So in the domain of physics. By identifying matter with spatial extension and by explaining the difference between one thing and another by recourse to the idea of movement communicated once and for all by God in a quantity which is constant, Descartes creates the conditions in which man will be able to foresee. All things are reduced to identity by defining them by the one characteristic attribute which they have in common, namely, extension; strict causality becomes assured by the immutability of God's action in a homogeneous world. In this way modern scientific experiment becomes a possibility: the laws which govern the physical world and which will continue to govern it to the end of all time may be discovered and used by man for his own ends. There is something paradoxical in the fact that, whereas scientific experiment plays only an ancillary role in Descartes' own practice – it exists merely to permit us to verify which of a number of equally probable conclusions arrived at deductively actually coincides with the facts; it intervenes at the end and not at the beginning of our inquiry – the profound significance of Cartesianism is precisely to give such a definition of the object of physics as to found the possibility of a science of laws reached through experiment.

Although today we are particularly sensitive to the anti-Aristotelian aspect of Cartesianism, contemporaries of Descartes had another opinion. Whilst recognizing in him a fellow mechanist, Roberval, Gassendi, Pascal and, as Lenoble has so well shown, Mersenne, were sensitive above all to the similarities between Descartes' approach to science and that of Aristotle. In other words, what Descartes had done was to adopt Aristotle's conception of physics as a demonstrative science based on necessary principles, whilst at the same time replacing the principles of Aristotle by those of his own finding. The objections raised against the *Meditations* stemmed largely from the fact that whereas Descartes, while rejecting the Aristotelian physics of quality, had not abandoned his logic, his contradictors had done both, and saw no necessity for the creation of a new form of dogmatism.

Descartes' great ambition, that of being the Aristotle of the modern age, was then never realized. He came too late, into a world which no longer had a place to offer for such enterprises, but the philosophic problems which he treated are still with us and the spirit in which he pursued his scientific quest still informs modern scientific thought. If the system of Descartes was a failure, Cartesianism as an attitude of mind was both fruitful and enduring.

DISCOURSE ON
THE METHOD OF PROPERLY
CONDUCTING ONE'S REASON AND
OF SEEKING THE TRUTH
IN THE SCIENCES

If this discourse appears too long to be read at one sitting, it may be split into six parts. In the first will be found various propositions concerning the sciences; in the second, the principal rules of the method which the author has sought out; in the third, some rules of moral conduct which he has derived from this method; in the fourth, the reasons by which he proves the existence of God and of the human soul, which are the basis of his metaphysics; in the fifth, the order of the questions in physics which he has sought to answer, and particularly the explanation of the movement of the heart and of some other difficulties peculiar to medicine, as also the difference between our soul and that of animals; and in the last section, the requirements he believes necessary in order to make further progress in research into natural phenomena and the reasons which have prompted him to write.

I

GOOD sense is the most evenly shared thing in the world, for each of us thinks he is so well endowed with it that even those who are the hardest to please in all other respects are not in the habit of wanting more than they have. It is unlikely that everyone is mistaken in this. It indicates rather that the capacity to judge correctly and to distinguish the true from the false, which is properly what one calls common sense or reason, is naturally equal in all men, and consequently that the diversity of our opinions does not spring from some of us being more able to reason than others, but only from our conducting our thoughts along different lines and not examining the same things. For it is not enough to have a good mind, rather the main thing is to apply it well. The greatest souls are capable of the greatest vices as well as of the greatest virtues, and those who go forward only very slowly can progress much further if they always keep to the right path, than those who run and wander off it.

For myself, I have never supposed that my mind was in any way out of the ordinary; indeed, I have often wished I could think as quickly and easily, have the same capacity for forming sharp and clear images, or a memory as rich and as ready to command as some. And I know of no other qualities than these which contribute to the perfection of the mind for, as far as reason or good sense is concerned, inasmuch as it is the only thing which makes us men and distinguishes us from the animals, I am ready to believe that it is complete and entire in each one of us, following in this the commonly held opinion of the philosophers who say that there are degrees only between *accidents* and not between the *forms* or *natures* of the *individuals* of a given *specie*.

But I shall not hesitate to say that I consider myself very fortunate to have found myself, from my early youth, on certain paths which led me to considerations and maxims out of which I have constructed a method which, I think, enables me gradually to increase my knowledge and to raise it little by little to the highest point which the mediocrity of my mind and the short span of my life will allow it to reach. For I have already reaped such a harvest from this method that, although in the assessment I make of myself, I try always to lean towards caution rather than to presumption, and although, looking at the various activities and enterprises of mankind with the eye of a philosopher, there is hardly one which does not seem to me vain and useless, I nevertheless feel extreme satisfaction at the progress which I think I have already made in the search for truth, and conceive such hopes for the future that if, among the activities of men as mere men, there is one which is thoroughly good and important, I dare to believe that it is the activity I have chosen.

However, I may be wrong, and perhaps I am mistaking a little copper and glass for gold and diamonds. I know how easily we can be mistaken in matters which concern us closely; and how much also the judgements of our friends must be suspect when they are in our favour. But I shall be very happy to reveal in this discourse the paths I have taken, and to present my life as in a picture, so that each may judge it, and so learning from what the public thinks of it, I may have a new means of instruction which I shall add to those which I am in the habit of using.

So my intention is not to teach here the method which everyone must follow if he is to conduct his reason correctly, but only to demonstrate how I have tried to conduct my own. Those who take the responsibility of giving precepts must think themselves more knowledgeable than those to whom they give them, and, if they make the slightest mistake, they are blameworthy. But, putting forward this essay as nothing more than an historical account, or, if you prefer, a fable in which, among certain examples one may follow, one will find also many others which it would be right not to copy, I hope it will be useful for some without being harmful to any, and that my frankness will be well received by all.

I was brought up from childhood on letters, and, because I had been led to believe that by this means one could acquire clear and positive knowledge of everything useful in life, I was extremely anxious to learn them. But, as soon as I had completed this whole course of study, at the end of which it is usual to be received into the ranks of the learned, I completely changed my opinion. For I was assailed by so many doubts and errors that the only profit I appeared to have drawn from trying to become educated, was progressively to have discovered my ignorance. And yet I was at one of the most famous schools in Europe, where I thought there must be learned men, if there were any such anywhere on earth. I had learnt there everything the others learned; and further, not contenting myself merely with the subjects taught, I had gone through all the books I could lay my hands on dealing with the occult and rare sciences. Moreover, I knew the assessments made of me by others, and it was obvious that they rated me no less than my fellow students, even though there were some among these who were already earmarked to succeed our teachers. And, finally, our century seemed just as flourishing and fertile in good minds as any earlier century. Consequently I took the liberty of judging all others by myself and of thinking that there was no body of knowledge in the world such as I had been led previously to believe existed.

I did not, however, fail to value the work we did in school. I knew that the languages one learns there are necessary for an understanding of the classics; that the grace of fables awakens the mind; that the memorable actions of history elevate it and that, if read with moderation and discernment, they help to form one's judgement; that to read good books is like holding a conversation with the most eminent minds of past centuries and, moreover, a studied conversation in which these authors reveal to us only the best of their thoughts; that oratory has incomparable power and beauty; that poetry has ravishing subtlety and sweetness; that mathematics contains some very ingenious inventions which can serve just as well to satisfy the curious as to make all arts and crafts easier and to lessen man's work; that writings which treat of ethics contain many very useful precepts and exhortations to virtue; that theology teaches how to gain Heaven; that philosophy gives the means by which one can speak plausibly on all matters and win the admiration of the less learned; that law, medicine and the other sciences bring honours and wealth to those who practise them; and finally that it is good to have examined them all, even those most full of superstition and falsehood, in order to know their true worth and to avoid being misled by them.

But I thought I had already given enough time to languages, and even also to the reading of the Ancients, to their histories and fables. For to converse with those of other centuries is almost the same as to travel. It is a good thing to know something of the customs and manners of various peoples in order to judge of our own more objectively and so not think everything which is contrary to our ways is ridiculous and irrational, as those who have seen nothing are in the habit of doing. But when one spends too much time travelling, one becomes eventually a stranger in one's own country; and when one is too interested in what went on in past centuries, one usually remains extremely ignorant of what is happening in this century. Moreover, fables make one imagine many events to be possible that are not, and even the most accurate historical accounts, while neither changing nor amplifying the value of things in order to make them more worth reading, at least almost always leave out the basest and least illustrious circumstances, with the result that what remains does not appear as it really is, and that those who base their own behaviour on the examples they draw from it risk falling into the extravagances of the *paladins* of our novels and conceive designs beyond their powers

I greatly esteemed eloquence and was in love with poetry, but I thought both were gifts of the mind rather than the fruit of study. Those with the most powerful reasoning, and who best order their thoughts to make them clear and intelligible, can always persuade us best of what they put forward even though they speak only the dialect of Lower Brittany and have never learnt rhetoric; and those who have the most pleasing conceits and know how to express them the most decoratively and harmoniously would always be the best poets, even though they knew nothing of the Art of Poetry.*

Above all I enjoyed mathematics, because of the certainty and self-evidence of its reasonings, but I did not yet see its true use and, thinking that it was useful only for the mechanical arts, I was astonished that on such firm and solid foundations nothing more exalted had been built, while on the other hand I compared the moral writings of the ancient pagans to the most proud and magnificent palaces built on nothing but sand and mud. They exalt the virtues and make them appear more estimable than anything in the world, but they do not sufficiently teach one to know them, and often what they give

* Reference to the *Ars Poetica* of Horace.

so fine a name to is only insensibility, or pride, or despair or parricide.

I revered our theology and aspired as much as anyone else to gain heaven; but having learnt as a certain fact that the path thither is open no less to the most ignorant than to the most learned, and that the revealed truths which lead to it are beyond our understanding, I would not have dared submit them to my weak powers of reasoning, and, in my opinion, to undertake the examination of them, and succeed, one would need some special grace from heaven and to be more than a mere man.

I shall say nothing about philosophy, except that, seeing that it has been cultivated by the very best minds which have ever existed over several centuries and that, nevertheless, not one of its problems is not subject to disagreement, and consequently is uncertain, I was not presumptuous enough to hope to succeed in it any better than others; and seeing how many different opinions are sustained by learned men about one item, without its being possible for more than one ever to be true, I took to be tantamount to false everything which was merely probable.

As for the other sciences, in so far as they borrow their principles from philosophy, I considered that nothing solid could have been built on such shifting foundations; and neither the honour nor the material gain held out by them was sufficient to induce me to study them. For I was not, thank Heaven, in the position of having to make a trade of knowledge to supplement my fortune; and although I did not adopt the attitude of the cynic, despising fame, I attached no importance to fame acquired undeservedly. Finally, as for the false sciences, I thought I knew sufficiently their worth already, not to be liable to be misled either by the promises of an alchemist or the predictions of an astrologer, the impostures of a magician, or the tricks or boasts of any of those who profess to know more than they do.

This is why, as soon as I reached an age which allowed me to emerge from the tutelage of my teachers, I abandoned the study of letters altogether, and resolving to study no other science than that which I could find within myself or else in the great book of the world, I spent the rest of my youth in travelling, seeing courts and armies, mixing with people of different humours and ranks, in gathering a varied experience, in testing myself in the situations which chance offered me, and everywhere reflecting upon whatever events I witnessed in such a way as to draw some profit from them. For it seemed to me that I might find much more truth in the reasonings which each one makes in matters that affect him closely, the result of which must be detrimental to him if his judgement is faulty, than from the speculations of a man of letters in his study which produce no concrete effect and which are of no other consequence to him except perhaps that the further away they are from common sense, the more vanity he will derive from them, because he will have had to use that much more skill and subtlety in order to try to make them seem dialectically probable. And I had always had an extreme desire to learn to distinguish true from false in order to see clearly into my own actions and to walk with safety in this life.

It is true that, while I merely observed the behaviour of others I found little basis in it for certainty, and I noticed almost as much diversity as I had done earlier among the opinions of philosophers. Hence the greatest profit I derived from it was that, seeing many things which, although they may seem to us very extravagant and ridiculous, are nevertheless commonly accepted and approved by other great peoples, I learned not to believe too firmly those things which I had been persuaded to accept by example and custom only; and in this way I freed myself gradually from many errors which obscure the natural light of our understanding and render us less capable of reason. But, after spending several years studying thus in the book of the world and seeking to gain experience, I resolved one day to study also myself and to use all the powers of my mind to choose the paths which I should follow. In this I was much more successful, it seems to me, than if I had never left either my country or my books.

I WAS, at that time, in Germany, whither the wars, which have not yet finished there, had called me, and as I was returning from the coronation of the Emperor to join the army, the onset of winter held me up in quarters in which, finding no company to distract me, and having, fortunately, no cares or passions to disturb me, I spent the whole day shut up in a room heated by an enclosed stove, where I had complete leisure to meditate on my own thoughts. Among these one of the first I examined was that often there is less perfection in works composed of several separate pieces and made by different masters, than in those at which only one person has worked. So it is that one sees that buildings undertaken and completed by a single architect are usually more beautiful and better ordered than those that several architects have tried to put into shape, making use of old walls which were built for other purposes. So it is that these old cities which, originally only villages, have become, through the passage of time, great towns, are usually so badly proportioned in comparison with those orderly towns which an engineer designs at will on some plain that, although the buildings, taken separately, often display as much art as those of the planned towns or even more, nevertheless, seeing how they are placed, with a big one here, a small one there, and how they cause the streets to bend and to be at different levels, one has the impression that they are more the product of chance than that of a human will operating according to reason. And if one considers that there have nevertheless always been officials responsible for the supervision of private building and for making it serve as an ornament for the public, one will see how difficult it is, by adding only to the constructions of others, to arrive at any great degree of perfection. So I thought to myself that the peoples who were formerly half savages, and who became civilized only gradually, making their laws only in so far as the harm done by crimes and quarrels forced them to do so, could not be so well organized as those who, from the moment at which they came together in association, observed the basic laws of some wise legislator; just as it is indeed certain that the state of the true religion, the laws of which God alone has made, must be incomparably better ordered than all the others. And, to speak of human things, I believe that, if Sparta greatly flourished in times past, it was not on account of the excellence of each of its laws taken individually, seeing that many were very strange and even contrary to good morals, but because, having been invented by one man only, they all tended towards the same end. And so I thought that the knowledge we acquire in books, at least that based on reasoning which is only probable and for which there is no proof, being composed and enlarged little by little by the opinions of many different people, does not approach the truth as closely as the simple reasoning of a man of good sense concerning things which he meets. So, finally, I thought that as we have all been children before being men, and that we have had to be governed for a long time by our appetites and our teachers, the ones being often in opposition to the others and neither perhaps always giving us the best advice, it is almost impossible that our judgements be as rational or as sound as they would have been if we had had the full use of our reason from the moment of our birth, and if we had never been guided by anything else.

It is true that we have no example of people demolishing all the houses in a town for the sole purpose of rebuilding them in a different way to make the streets more beautiful; but one does see many people knock down their own in order to rebuild them, and that even in some cases they have to do this because the houses are in danger of falling down and the foundations are insecure. With this example in mind, I felt convinced that it would be unreasonable for an individual to conceive the plan of reforming a State by changing everything from the foundations up and by overthrowing it in order to set it up again, or even to reform the body of the sciences or the order established in our schools for teaching it, but that, on the other hand, as far as all the opinions I had accepted hitherto were concerned, I could not do better than undertake once and for all to be rid of them in order to replace them afterwards either by better ones, or even by the same, once I had adjusted them by the plumb-line of reason. And I firmly believed that, by this means, I would succeed in ordering my life much better than if I built only on old foundations and leaned on principles inculcated in me in youth without having ever examined them to see if they were true. For, although I could see several difficulties in this course, they were not all totally irremediable, nor are they comparable to those which arise in the reformation of the least things affecting the State. These great bodies are too difficult to raise up again, once knocked down, or even to hold up, once shaken, and their fall can only be very heavy. Then, as for their imperfections, if they have any, and the mere diversity among them suffices to assure us that many of them have imperfections, usage had doubtless softened many of them considerably, and has even insensibly averted or corrected many which one could not have so well remedied by prudence. Finally, these imperfections are almost always more bearable than changing them would be, in the same way that the high roads which wind round between mountains become gradually so smooth and convenient by dint of being much used, that it is much better to follow them than to undertake to go more directly by scrambling up rocks and going down to the very foot of precipices.

This is why I can never bring myself to approve those meddling and restless spirits, who, being called neither by birth nor by fortune to administer public affairs, are forever imagining some reform of the State. And if I thought there was the least thing in this discourse by which anyone could suspect me of this madness, I would be very sorry to allow it to be published. My plan has never gone beyond trying to reform my own thoughts and to build on a foundation which is wholly my own. And if, my work having sufficiently satisfied me, I set it out here as a model for you, it is not on this account that I would advise anyone to copy it. Those whom God has better endowed with his blessings will perhaps have more elevated designs, but I very much fear that even mine be too bold for many people. The mere resolve to divest oneself of all one's former opinions is not an example to be followed by everyone; and the world is made up almost entirely of two types of mind for which this would not be at all suitable: namely, those who, thinking themselves to be cleverer than they are, cannot help judging prematurely and do not have the patience to conduct their thought in an orderly way, with the result that, once they have taken the liberty of doubting accepted principles and of leaving the common path, they would never be able to keep to the road one must take in order to go straight forward and would remain lost all their lives; secondly, there are those who, having enough sense or modesty to know that they are less able to distinguish the true from the false than some from whom they can learn, ought rather to content themselves with following the opinions of these others instead of seeking better opinions themselves.

And as for me, I would undoubtedly have been in this second category if I had had only one master, or if I had never known the differences of opinion that there have always been among the most learned. But, having learnt from the time I was at school that there is nothing one can imagine so strange or so unbelievable that has not been said by one or other of the philosophers; and since then, while travelling, having recognized that all those who hold opinions quite opposed to ours are not on that account barbarians or savages, but that many exercise as much reason as we do, or more; and having considered how a given man, with his given mind, being brought up from childhood among the French or Germans, becomes different from what he would be if he had always lived among the Chinese or among cannibals; and how, down to our very fashions in dress, what pleased us ten years ago and will perhaps please us again before another ten years are out, now seems to us extravagant and laughable, I was convinced that our beliefs are based much more on custom and example than on any certain knowledge, and, nevertheless that the assent of many voices is not a valid proof for truths which are rather difficult to discover, because they are much more likely to be found by one single man than by a whole people. Thus I could not choose anyone whose opinions it seemed to me I ought to prefer to those of others, and I found myself constrained, as it were, to undertake my own guidance.

But, like a man who walks alone, and in the dark, I resolved to go so slowly, and to use such caution in all things that, even if I went forward only very little, I would at least avoid falling. Moreover, I did not wish to begin to reject completely any of the opinions which might have slipped earlier into my mind without having been introduced by reason, until I had first given myself enough time to make a plan of the work I was undertaking, and to seek the true method of arriving at knowledge of everything my mind was capable of grasping.

When I was younger, I had studied a little logic in philosophy, and geometrical analysis and algebra in mathematics, three arts or sciences which would appear apt to contribute something towards my plan. But on examining them, I saw that, regarding logic, its syllogisms and most of its other precepts serve more to explain to others what one already knows, or even, like the art of Lully, to speak without judgement of those things one does not know, than to learn anything new. And although logic indeed contains many very true and sound precepts, there are, at the same time, so many others mixed up with them, which are either harmful or superfluous, that it is almost as difficult to separate them as to extract a Diana or a Minerva from a block of unprepared marble. Then, as for the geometrical analysis of the ancients and the algebra of the moderns, besides the fact that they extend only to very abstract matters which seem to be of no practical use, the former is always so tied to the inspection of figures that it cannot exercise the understanding without greatly tiring the imagination, while, in the latter, one is so subjected to certain rules and numbers that it has become a confused and obscure art which oppresses the mind instead of being a science which cultivates it. This was why I thought I must seek some other method which, while continuing the advantages of these three, was free from their defects. And as a multiplicity of laws often furnishes excuses for vice, so that a State is much better ordered when, having only very few laws, they are very strictly observed, so, instead of this great number of precepts of which logic is composed, I believed I would have sufficient in the four following rules, so long as I took a firm and constant resolve never once to fail to observe them.

The first was never to accept anything as true that I did not know to be evidently so: that is to say, carefully to avoid precipitancy and prejudice, and to include in my judgements nothing more than what presented itself so clearly and so distinctly to my mind that I might have no occasion to place it in doubt.

The second, to divide each of the difficulties that I was examining into as many parts as might be possible and necessary in order best to solve it.

The third, to conduct my thoughts in an orderly way, beginning with the simplest objects and the easiest to know, in order to climb gradually, as by degrees, as far as the knowledge of the most complex, and even supposing some order among those objects which do not precede each other naturally.

And the last, everywhere to make such complete enumerations and such general reviews that I would be sure to have omitted nothing.

These long chains of reasonings, quite simple and easy, which geometers are accustomed to using to teach their most difficult demonstrations, had given me cause to imagine that everything which can be encompassed by man's knowledge is linked in the same way, and that, provided only that one abstains from accepting any for true which is not true, and that one always keeps the right order for one thing to be deduced from that which precedes it, there can be nothing so distant that one does not reach it eventually, or so hidden that one cannot discover it. And I was in no great difficulty in seeking which to begin with because I knew already that it was with the simplest and easiest to know; and considering that, among all those who have already sought truth in the sciences, only the mathematicians have been able to arrive at any proofs, that is to say, certain and evident reasons, I had no doubt that it was by the same things which they had examined that I should begin, although I did not expect any other usefulness from this than to accustom my mind to nourish itself on truths and not to be content with false reasons. But it was not my intention, to this end, to try to learn all those particular sciences which come together under the name of mathematics, and, seeing that, even though their objects are different, they are all concordant in so far as they consider only the various relations or proportions to be found among these objects, I thought it would be best if I were to examine only these proportions in general and without supposing their existence except in those things which would serve to make my knowledge of them easier, and also without in any way binding them to these things in order to be able to apply them so much the better afterwards to all the others to which they might be applicable. Then, having taken note that, in order to know them, I would need sometimes to consider each one individually, and sometimes only to remember them, or to consider several of them all together, I thought that to consider them best individually I should represent them as straight lines because I could find nothing simpler and nothing which I could grasp by my imagination and my senses more clearly, but for the purpose of remembering them or taking several together, I should have to designate them by certain symbols, the briefest possible, and by this means, I would borrow all the best from geometric analysis and from algebra, and would correct all the defects of the one by the other.

And, indeed, I dare say that the exact observation of these few precepts that I had chosen gave me such ease in unravelling all the questions covered by these two sciences in the two or three months which I spent on examining them, having begun by the simplest and most general, and each truth that I found being a rule which served afterwards to find others, not only did I resolve several questions which I had earlier judged to be very difficult, but it also seemed to me, towards the end, that I could determine, even in those of the solution of which I was ignorant, by what means and how far it would be possible to resolve them. In this I shall not perhaps appear to you to be too vain, if you consider that, as there is only one truth of each thing, whoever finds it knows as much about the thing as there is to be known, and that, for example, a child who has been taught arithmetic, having added up according to the rules, can be sure that he has found out, as far as the sum he was examining is concerned, all that the human mind is capable of finding out. For, after all, the method which teaches one to follow the true order and to enumerate exactly all the factors required for the solution of a problem, contains everything which gives certainty to the rules of arithmetic.

But what satisfied me the most about this method was that, through it, I was assured of using my reason in everything, if not perfectly, at least to the best of my ability. Moreover, I felt that, in practising it, my mind was accustoming itself little by little to conceive its objects more clearly and distinctly, and not having subjected it to any particular matter, I promised myself that I would apply it just as usefully to the difficulties of the other sciences as I had to those of algebra. Not that, for all that, I dared undertake straight away to examine all those which presented themselves, for this itself would have been contrary to the order which my method prescribes. But having taken note that their principles must all be borrowed from philosophy, in which I still found no assured principles, I thought that I must first of all try to establish some, and that, this being the most important thing in the world, and one in which undue haste and prejudice were to be feared the most, I ought not to undertake this task until I had achieved a much more mature age than the twenty-three years I was then, and until I had spent a considerable period beforehand in preparing myself, as much in rooting out of my mind all the false opinions that had been instilled into me earlier, as in amassing much experience to serve afterwards as the matter of my reasonings, and in always practising my prescribed method so as to strengthen myself in it more and more.

IN the same way as it is not sufficient, before beginning to rebuild the house in which one lives, only to pull it down and to provide material and architects, or oneself to try one's hand at architecture, and moreover, to have drawn the plan carefully, but one must also provide oneself with some other accommodation in which to be lodged conveniently while the work is going on, so, also, in order that I might not remain irresolute in my actions during the time that my reason would oblige me to be so in my judgements, and so that I would not cease to live from that time forward as happily as I could, I formed a provisional moral code which consisted of only three or four maxims, which I am willing to disclose.

The first was to obey the laws and customs of my country, firmly preserving the religion into which God was good enough to have me instructed from childhood, and governing myself in all other matters according to the most moderate opinions and those furthest from excess, commonly accepted in practice by the most prudent people with whom I should have to live. For, beginning already to discount my own opinions because I intended to submit them all to examination, I was assured I could not do better than to follow those of the most prudent. And although there may be perhaps as sensible people among the Persians or the Chinese as among ourselves, it seemed to me that the most useful would be to adapt my behaviour to that of those with whom I would have to live, and that, to know which were truly their opinions, I should take notice rather of what they practised than of what they said, not only because in the corruption of our manners there are few people who are willing to say all they believe, but also because many do not themselves know what they believe; for the action of thought by which one believes a thing, being different from that by which one knows that one believes it, they often exist the one without the other. And among many opinions equally accepted I would choose only the most moderate, as these are always the most easily applicable in practice, and very likely the best, all excess being usually bad; moreover, I would wander less far from the true path if I should prove to have taken the wrong opinion, than if, having chosen one of the extremes, it were the other extreme that I ought to have followed. In particular, I put among the excesses all the promises by which some part of one's freedom is taken away. Not that I disapproved laws which, to counteract the inconstancy of weak minds, permit, when one has some pious design or, in the case of the security of commerce, a merely secular design, that one should make vows or contracts which oblige one to persevere in them. But, because I saw in the world nothing which remained always in the same state, and as, in my own particular case, I promised myself to make my judgements more and more perfect and not to make them worse, I would have thought I had committed a grave error against reason if, because I once approved something, I felt obliged to consider it good later, when it had perhaps ceased to be good or I had ceased to consider it so.

My second maxim was to be as firm and resolute in my actions as I could, and to follow no less constantly the most doubtful opinions, once I had determined on them, than I would if they were very assured, imitating in this travellers, who, finding themselves astray in some forest, must not wander, turning now this way now that, and even less stop in one place, but must walk always as straight as they can in a given direction, and not change direction for weak reasons, even though it was perhaps only chance in the first place which made them choose it; for, by this means, if they do not go exactly where they wish to go they will arrive at least somewhere in the end where they will very likely be better off than in the middle of a forest. And so it is that, the actions of life often brooking no delay, it is a certain truth that, when we are powerless to discern the truest opinions, we must follow the most probable, and although we see no more probability in some than in others, we must nevertheless settle on some and consider them afterwards no longer as being doubtful, in so far as they relate to practice, but as very true and very certain, because the reason which has caused us to settle upon them is itself such. And this was capable from then on of freeing me from all the repentance and remorse which habitually agitate the consciences of those weak and wavering minds which allow themselves to proceed with vacillation to practise as being good things which they judge afterwards to be bad.

My third maxim was to try always to conquer myself rather than fortune, and to change my desires rather than the order of the world, and generally to accustom myself to believing that there is nothing entirely in our power except our thoughts, so that after we have done our best regarding things external to us, everything in which we do not succeed is for us absolutely impossible. And this alone seemed to me to be sufficient to prevent me from desiring anything in the future that I could not obtain, and therefore to make me contented; for, our will naturally tending only to wish for things which our understanding tells it are in some way attainable, it is certain that, if we consider all external goods as being equally beyond our power, we should have no more regret for lacking those which seem to be due to us by birth, when we are deprived of them without any fault on our part, than we have for not possessing the kingdoms of China or Mexico; and thus making, as the saying is, a virtue of necessity, we shall no more wish to be well when we are ill, or to be free when we are in prison, than we do now to have bodies made of some material as incorruptible as diamonds, or wings to fly like birds. But I admit that long practice and often repeated meditation are necessary to accustom oneself to look at everything in this way, and I think it is in this that lies the secret of those philosophers who in former times were able to escape the sway of Fortune and, in spite of suffering and poverty, to rival their gods in happiness. For, occupied ceaselessly in considering the limits laid on them by nature, they convinced themselves so perfectly that nothing was in their power except their thoughts, that this alone was enough to stop them having any longing for other things; and they possessed their thoughts so absolutely that they had thereby some reason to consider themselves richer, more powerful, freer and happier than any other men, who, if not having this philosophy, however favoured by nature and fortune they might be, never have this control over the desires of their will.

Finally, to conclude this code of morals, I thought it advisable to examine minutely the diverse occupations of men in this life, to try to choose the best, and, without wishing to comment on the occupations of others, I thought I could not do better than to continue in my present one, that is to say, to devote all my life to the cultivation of my reason, and to progress as much as possible in the knowledge of truth, following the method I had prescribed for myself. I had experienced such extreme satisfaction since beginning to make use of this method that I did not think it possible to experience gentler or more innocent joys in this life; and discovering daily, by its means, truths which seemed of some importance and generally unknown to other men, the satisfaction

I obtained from it so filled my mind that nothing else mattered to me. Besides, the three foregoing maxims were based only on the design I had to continue to learn. For, God having given each of us some light of reason to discern true from false, I could not have believed I ought to content myself for one moment with the opinions of others, if I had not intended to use my own judgement to examine them in due course, and I would not have been able to avoid having scruples about following them, if I had not hoped to lose no opportunity thereby of finding better ones if such existed. And finally, I would not have been able to limit my desires or to be happy, if I had not followed a path by which, thinking I was certain of acquiring all the knowledge of which I was capable, I thought also that I was certain, by the same means, of acquiring all the true goods that would ever be in my power. Since our will tends neither to follow nor to flee anything except in so far as our understanding represents it as good or bad, it suffices to judge well in order to act well, and to judge to the best of one's ability, in order also to do one's best, that is to say, in order to acquire all the virtues and with them all the other goods one is capable of acquiring, and when one is sure that this situation exists, one cannot fail to be happy.

Having thus provided myself with these maxims, and having put them on one side along with the truths of faith, which have always held first place in my belief, I judged that, as for the rest of my opinions, I was free to undertake to divest myself of them. And inasmuch as I hoped to be able to complete this task by frequenting other men rather than by remaining any longer shut up in the stove-heated room in which I had had all these thoughts, winter was not yet fully over before I took to travelling again. And through all the nine years which followed I did nothing but wander here and there in the world, trying to be spectator rather than actor in all the comedies which were being played there; and reflecting particularly in each matter on what might render it doubtful and give occasion for error, I rooted out from my mind, during this time, all the errors which had introduced themselves into it hitherto. Not that, in so doing, I imitated the sceptics who doubt only for doubting's sake, and affect to be always undecided; for, on the contrary, my whole plan had for its aim assurance and the rejection of shifting ground and sand in order to find rock or clay. This object I achieved well enough, it seems to me, inasmuch as, trying to discover the falseness or uncertainty of the propositions I examined, not with weak guesses, but with clear and assured reasonings, I found none so doubtful that I could not draw from it some sufficiently certain conclusion, even if this might not be other than that the proposition contained nothing certain. And as, in knocking down an old house, one usually reserves the demolished material to be used for building a new one, so in destroying all those of my opinions that I judged to be ill-founded, I made various observations and acquired several experiences which have served me since to establish more certain ones. And, in addition, I continued to practise the method I had laid down for myself; for, besides taking care generally to conduct all my thoughts according to its rules, I put some time aside, now and again, which I used in particular to exercise the method in the solution of mathematical difficulties, or even in that of some others which I could make almost like mathematical problems by separating them from all the principles of the other sciences, which I did not find solid enough, as you will see I have done in the case of several which are explained in this volume. And so, without living outwardly any differently from those, who, having no occupation other than to lead an agreeable and innocent life, are careful not to confuse pleasures and vices, and who, to enjoy their leisure without boredom indulge in all honourable distractions, I continued to follow my plan and to progress in the knowledge of truth, perhaps more than if I had done nothing but read books or spent my time in the company of men of letters.

However, these nine years passed before I had as yet made any definitive judgement regarding the difficulties which the learned are accustomed to debate, or had begun to seek the bases of any philosophy more certain than the vulgar. And the example of many excellent minds which, having had this plan previously, did not appear to me to have succeeded, made me think it was so difficult, that I would not have dared to undertake it so early if I had not seen that some people were already putting about the rumour that I had completed my task. I cannot say on what they based this opinion; and if I contributed anything towards this rumour by anything I said, it must have been by admitting what I did not know, more naïvely than is usual among those who have studied a little, and also perhaps by displaying the reasons I had for doubting many things that others accepted as certain, rather than by boasting of any knowledge. But, being proud enough not to want anyone to take me for something different from what I was, I thought I ought to try by all means to make myself worthy of the reputation I was being given; and it is exactly eight years since this wish made me decide to leave all those places where I had acquaintances, and to withdraw here to a country where the long duration of the war has established such discipline that the armies maintained there seem to serve only to ensure that the fruits of peace are enjoyed with the maximum of security; and where, in the midst of a great crowd of busy people, more concerned with their own business than curious about that of others, without lacking any of the conveniences offered by the most populous cities, I have been able to live as solitary and withdrawn as I would in the most remote of deserts.

I DO not know if I ought to tell you about the first meditations I pursued there, for they are so abstract and unusual that they will probably not be to the taste of everyone; and yet, so that one may judge if the foundations I have laid are firm enough, I find myself to some extent forced to speak of them. I had long ago noticed that, in matters relating to conduct, one needs sometimes to follow, just as if they were absolutely indubitable, opinions one knows to be very unsure, as has been said above; but as I wanted to concentrate solely on the search for truth, I thought I ought to do just the opposite, and reject as being absolutely false everything in which I could suppose the slightest reason for doubt, in order to see if there did not remain after that anything in my belief which was entirely indubitable. So, because our senses sometimes play us false, I decided to suppose that there was nothing at all which was such as they cause us to imagine it; and because there are men who make mistakes in reasoning, even with the simplest geometrical matters, and make paralogisms, judging that I was as liable to error as anyone else, I rejected as being false all the reasonings I had hitherto accepted as proofs. And finally, considering that all the same thoughts that we have when we are awake can also come to us when we are asleep, without any one of them then being true, I resolved to pretend that nothing which had ever entered my mind was any more true than the illusions of my dreams. But immediately afterwards I became aware that, while I decided thus to think that everything was false, it followed necessarily that I who thought thus must be something; and observing that this truth: *I think, therefore I am*, was so certain and so evident that all the most extravagant suppositions of the sceptics were not capable of shaking it, I judged that I could accept it without scruple as the first principle of the philosophy I was seeking.

Then, examining attentively what I was, and seeing that I could pretend that I had no body and that there was no world or place that I was in, but that I could not, for all that, pretend that I did not exist, and that, on the contrary, from the very fact that I thought of doubting the truth of other things, it followed very evidently and very certainly that I existed; while, on the other hand, if I had only ceased to think, although all the rest of what I had ever imagined had been true, I would have had no reason to believe that I existed; I thereby concluded that I was a substance, of which the whole essence or nature consists in thinking, and which, in order to exist, needs no place and depends on no material thing; so that this 'I', that is to say, the mind, by which I am what I am, is entirely distinct from the body, and even that it is easier to know than the body, and moreover, that even if the body were not, it would not cease to be all that it is.

After this, I considered in general what is needed for a proposition to be true and certain; for, since I had just found one which I knew to be so, I thought that I ought also to know what this certainty consisted of. And having noticed that there is nothing at all in this, *I think, therefore I am*, which assures me that I am speaking the truth, except that I see very clearly that in order to think one must exist, I judged that I could take it to be a general rule that the things we conceive very clearly and very distinctly are all true, but that there is nevertheless some difficulty in being able to recognize for certain which are the things we see distinctly.

Following this, reflecting on the fact that I had doubts, and that consequently my being was not completely perfect, for I saw clearly that it was a greater perfection to know than to doubt, I decided to inquire whence I had learned to think of some thing more perfect than myself; and I clearly recognized that this must have been from some nature which was in fact more perfect. As for the notions I had of several other things outside myself, such as the sky, the earth, light, heat and a thousand others, I had not the same concern to know their source, because, seeing nothing in them which seemed to make them superior to myself, I could believe that, if they were true, they were dependencies of my nature, in as much as it had some perfection; and, if they were not, that I held them from nothing, that is to say that they were in me because of an imperfection in my nature. But I could not make the same judgement concerning the idea of a being more perfect than myself; for to hold it from nothing was something manifestly impossible; and because it is no less contradictory that the more perfect should proceed from and depend on the less perfect, than it is that something should emerge out of nothing, I could not hold it from myself; with the result that it remained that it must have been put into me by a being whose nature was truly more perfect than mine and which even had in itself all the perfections of which I could have any idea, that is to say, in a single word, which was God. To which I added that, since I knew some perfections that I did not have, I was not the only being which existed (I shall freely use here, with your permission, the terms of the School) but that there must of necessity be another more perfect, upon whom I depended, and from whom I had acquired all I had; for, if I had been alone and independent of all other, so as to have had from myself this small portion of perfection that I had by participation in the perfection of God, I could have given myself, by the same reason, all the remainder of perfection that I knew myself to lack, and thus to be myself infinite, eternal, immutable, omniscient, all-powerful, and finally to have all the perfections that I could observe to be in God. For, consequentially upon the reasonings by which I had proved the existence of God, in order to understand the nature of God as far as my own nature was capable of doing, I had only to consider, concerning all the things of which I found in myself some idea, whether it was a perfection or not to have them: and I was assured that none of those which indicated some imperfection was in him, but that all the others were. So I saw that doubt, inconstancy, sadness and similar things could not be in him, seeing that I myself would have been very pleased to be free from them. Then, further, I had ideas of many sensible and bodily things; for even supposing that I was dreaming, and that everything I saw or imagined was false, I could not, nevertheless, deny that the ideas were really in my thoughts. But, because I had already recognized in myself very clearly that intelligent nature is distinct from the corporeal, considering that all composition is evidence of dependency, and that dependency is manifestly a defect, I thence judged that it could not be a perfection in God to be composed of these two natures, and that, consequently, he was not so composed; but that, if there were any bodies in the world or any intelligences or other natures which were not wholly perfect, their existence must depend on his power, in such a way that they could not subsist without him for a single instant.

I set out after that to seek other truths; and turning to the object of the geometers, which I conceived as a continuous body, or a space extended indefinitely in length, width and height or depth, divisible into various parts, which could have various figures and sizes and be moved or transposed in all sorts of ways – for the geometers take all that to be in the object of their study – I went through some of their simplest proofs. And having observed that the great certainty that everyone attributes to them is based only on the fact that they are clearly conceived according to the rule I spoke of earlier, I noticed also that they had nothing at all in them which might assure me of the existence of their object. Thus, for example, I very well perceived that, supposing a triangle to be given, its three angles must be equal to two right angles, but I saw nothing, for all that, which assured me that any such triangle existed in the world; whereas, reverting to the examination of the idea I had of a perfect Being, I found that existence was comprised in the idea in the same way that the equality of the three angles of a triangle to two right angles is comprised in the idea of a triangle or, as in the idea of a sphere, the fact that all its parts are equidistant from its centre, or even more obviously so; and that consequently it is at least as certain that God, who is this perfect Being, is, or exists, as any geometric demonstration can be.

But what persuades many people that it is difficult to know this, and even also to know what their soul is, is that they never lift their minds above tangible things, and that they are so accustomed not to think of anything except by imagining it, which is a mode of thinking peculiar to material objects, that everything which is not within the realm of imagination seems to them unintelligible. This is evident enough from the fact that even the philosophers hold as a maxim in the Schools, that there is nothing in the understanding which has not first been in the senses, in which, however, it is certain that ideas about God and the soul have never been; and it seems to me that those who wish to use their imagination to understand them are doing just the same as if, to hear sounds or smell odours, they attempted to use their eyes; except that there is still this difference, that the sense of sight assures us no less of the truth of its objects than do the senses of smell and hearing, whereas neither our imagination nor our senses could ever assure us of anything, if our understanding did not intervene. Finally, if there are still men who are not sufficiently persuaded of the existence of God and of their soul by the reasons I have given, I would like them to know that all the other things of which they think themselves perhaps more assured, such as having a body, and that there are stars and an earth, and such like, are less certain; for, although we may have a moral assurance of these things, which is such that it seems that, short of being foolish, no one can doubt their existence, at the same time also, short of being unreasonable, when it is a question of a metaphysical certainty, one cannot deny that there are not sufficient grounds for being absolutely assured, when one observes that one can in the same way imagine, being asleep, that one has another body, and that one sees other stars and another earth, without there being anything of the sort. For how does one know that the thoughts which come while one dreams are false rather than the others, seeing that they are often no less strong and clear? And may the most intelligent men study this question as much as they please, I do not believe that they can give any reason which would be sufficient to remove this doubt, unless they presuppose the existence of God. For, firstly, even the rule which I stated above that I held, namely, that the things we grasp very clearly and very distinctly are all true, is assured only because God is or exists, and because he is a perfect Being, and because everything that is in us comes from him; whence it follows that our ideas and notions, being real things and coming from God, in so far as they are clear and distinct, cannot to this extent be other than true. Accordingly, if we often enough have ideas which contain errors, they can only be those which contain something confused and obscure, because in this they participate in nothingness, that is to say that they are in us in this confused way only because we are not completely perfect. And it is evident that it is no less contradictory that error or imperfection, as such, should proceed from God, than that truth or perfection should come from nothingness. But, if we did not know that all that is in us which is real and true comes from a perfect and infinite Being, we would have no reason which would assure us that, however clear and distinct our ideas might be, they had the perfection of being true.

But, after knowledge of God and of the soul has thus made us certain of this rule, it is a simple matter to understand that the dreams we imagine when we are asleep should not in any way make us doubt the truth of the thoughts we have when we are awake. For, even if it should happen that, while sleeping, one should have some quite distinct idea, as, for example, if a geometer were to discover some new demonstration, his being asleep would not prevent it from being true; and as for the most ordinary error of our dreams, which consists in representing to us various objects in the same way as our waking senses do, it does not matter that they give us occasion to doubt the truth of such ideas, because they can also lead us into error often enough without our being asleep, as when those who have jaundice see everything yellow, or when the stars or other very distant bodies seem to us much smaller than they are. For, finally, whether we are awake or asleep, we should never let ourselves be persuaded except on the evidence of our reason. And it is to be observed that I say: of our reason, and not: of our imagination or our senses. For, although we see the sun very clearly, we should not on that account judge that it is only as large as we see it; and we can well imagine distinctly a lion's head grafted on to the body of a goat, without concluding on that account that there is any such chimera in the world; for reason does not dictate that what we see or imagine thus is true, but it does tell us that all our ideas and notions must have some basis in truth, for it would not be possible that God, who is all perfect and true, should have put them in us unless it were so. And because our reasonings are never so clear or complete while we sleep as when we are awake, even though sometimes our imaginations are as vivid and distinct or even more so, reason tells us that, it not being possible that our thoughts should all be true, because we are not absolutely perfect, what truth there is in them will undoubtedly be found in those we have when we are awake rather than in those we have in our dreams.

I SHOULD be very pleased to continue, and to show here the complete chain of the other truths that I deduced from these first ones; but as in order to do this it would now be necessary for me to speak of several questions about which the philosophers with whom I have no wish to embroil myself, are in dispute, I believe it will be better for me to abstain, and mention them only in general terms, in order to leave the more judicious to decide whether it would be useful that the public were informed of them in greater detail. I have always remained firm in the resolution I made not to suppose the existence of any other principle than that which I have just used to demonstrate the existence of God and of the soul, and not to accept anything as being true which did not seem to me more clear and certain than had previously the demonstrations of the geometers; and nevertheless, I dare to say that, not only have I found the means to satisfy myself in a short space of time, concerning all the principal difficulties which one usually treats of in philosophy, but also I have observed certain laws which God has so established in nature and of which he has impressed such notions in our souls, that having reflected on them sufficiently, we cannot be in any doubt that they are strictly observed in everything which exists or which happens in the world. Then, by considering the series of these laws, it appears to me that I have discovered many truths more useful and more important than anything I had learned before or even hoped to learn.

But, because I have tried to explain the most important of these in a treatise which a number of considerations prevent me from publishing, I could not make them better known than by saying here briefly what the treatise contains. It was my plan to include in it everything I thought I knew, before writing it down, concerning the nature of material things. But, in the same way that painters, being unable to represent equally well in a flat picture all the various faces of a solid body, choose one of the principal ones which they place alone in the light, and, putting the others in the shade, let them appear only in so far as they can be seen while one is looking at the principal one, so, fearing that I would not be able to put into my discourse all that I had in my mind, I undertook only to expound fully what I understood about light; then, to take the opportunity to add to it something on the sun and the fixed stars, because light almost wholly stems from them; something about the firmament, because it transmits light; about the planets, comets and earth, because they reflect it; and in particular about all the bodies on earth, because they are either coloured, or transparent or luminous; and finally about man, because he sees all this. And in order to put all these new truths in a less crude light and to be able to say more freely what I think about them, without being obliged to accept or to refute what are accepted opinions among the philosophers and theologians, I resolved to leave all these people to their disputes, and to speak only of what would happen in a new world, if God were now to create, somewhere in imaginary space, enough matter to compose it, and if he were to agitate diversely and confusedly the different parts of this matter, so that he created a chaos as disordered as the poets could ever imagine, and afterwards did no more than to lend his usual preserving action to nature, and let her act according to his established laws. So, firstly, I described this substance, and tried to represent it such that, to my mind, there is nothing in the world clearer or more readily understandable, except what has already been said about God and the soul; for I even supposed expressly that there was in it none of those forms or qualities about which there is disagreement in the Schools, nor generally anything the knowledge of which was not so natural to our minds that one could not even pretend to be ignorant of it. Moreover, I showed what are the laws of nature and, without basing my reasonings on any other principle than on the infinite perfections of God, I tried to prove all those about which one might have had some doubt and to show that they are such that, if God had created many worlds, there could be none in which they failed to be observed. After this, I showed how most of the matter of this chaos must, in accordance with these laws, dispose and arrange itself in a certain way which would make it similar to our skies; how, in the meantime, some particles must compose an earth, others planets and comets, and still others a sun and fixed stars. And here, enlarging upon the subject of light, I explained at some length the nature of the light which is found in the sun and the stars, and how, from them, it crosses in an instant the immense expanses of the heavens, and how it is reflected from the planets and comets towards the earth. I added also many things about the substances, position, movements and all the different qualities of these heavens and stars; so that I thought I had said enough about them to show that there is nothing to be seen in those of this world which should not, or at least could not, appear just the same in those of the world I was describing. Next I came to speak of the earth in particular and to point out how it is that, although I had expressly supposed that God had put no weight in the matter of which it is composed, all its parts do nonetheless tend exactly towards its centre; how, there being water and air on its surface, the disposition of the heavens and the heavenly bodies, principally of the moon, must cause an ebb and flow in all respects similar to that which we see in our seas, and furthermore a certain current, as much of water as of air, from east to west, such as one notices between the tropics; how the mountains, seas, springs and rivers could form themselves naturally, and metals appear in the mines, and plants grow in the countryside, and in general, how all the bodies one calls mixed or composite could be engendered. And, among other things, because, apart from the stars, I knew nothing in the world except fire which produces light, I strove to make clearly understood everything belonging to its nature, how it is made, how it is fed, how sometimes there is only heat without light, and sometimes only light without heat; how it can introduce different colours into different bodies, and diverse other qualities; how it melts some things and hardens others; how it can consume almost everything or convert into ashes and smoke; and finally how from these ashes, by the mere power of its action, it forms glass; for as this transmutation of ashes into glass seems to me as wonderful as any other which happens in nature, I took particular pleasure in describing it.

However, I did not wish to infer from all these things that this world had been created in the way I described for it is very much more likely that, from the beginning, God made it as it was to be. But it is certain, and this is an opinion commonly held among the theologians, that the action by which he conserves it now is the same as that by which he created it; so that even though he did not at the beginning give it any other form than that of chaos, provided that he had established the laws of nature and lent it his preserving action to allow it to act as it does customarily, one can believe, without discrediting the miracle of creation, that in this way alone, all things which are purely material could in time have made themselves such as we see them today; and their nature is much easier to grasp when one sees them being fully made from the start.

From the description of inanimate objects and of plants I passed to that of animals, and particularly of man. But, because I had not yet enough knowledge to talk about it in the same way as the rest, that is to say, in proving effects by causes, and demonstrating from what elements and in what way nature must produce them, I contented myself with supposing that God made the body of man entirely similar to one of ours, as much in the outward shape of the members as in the internal conformation of the organs, without making it of a different matter from the one which I have described, and without putting in it at the beginning any rational soul, or any other thing to serve as a vegetative or sensitive soul, but merely kindling in his heart one of those fires without light that I had already explained and which I did not conceive as being of any other nature than that which heats hay when it has been stacked before it is dry, or which makes new wine ferment, when it is left to ferment on the lees. For, examining the functions which could, consequentially, be in this body, I found precisely all those which can be in us without our thinking of them, and therefore, without our soul, that is to say, that part distinct from the body about which it has been said above that its nature is only to think, contributing to them, and these are all the same functions in which one can say that the animals, devoid of reason, resemble us. But I was unable for all that to find any of those functions which, being dependent on thought, are the only ones which belong to us as men, whereas I found them all afterwards, once I had supposed that God created a rational soul, and joined it to this body in a particular way which I described.

But, so that one can see how I treated this matter there, I wish to give here the explanation of the movement of the heart and arteries, from which, being the first and most general that one observes in animals, one will be able to judge easily what one should think of all the others. And, so that one should have less difficulty in understanding what I shall say about it, I would like those who are not versed in anatomy to take the trouble, before reading this, to have cut open in front of them the heart of some large animal which has lungs, because it is, in all of them, similar enough to that of man, and to be shown its two ventricles or cavities. Firstly that in the right side, to which correspond two very large tubes: i.e. the *vena cava* (hollow vein), which is the main receptacle of the blood and is like the trunk of a tree of which all the other veins of the body are the branches; and the *vena arteriosa* (arterial vein) which is badly named thus because it is in fact an artery, which, originating in the heart, divides, having come out of the heart, into many branches which spread out throughout the lungs. Then the cavity in the left side, which similarly has two pipes which are in width equal to or larger than the preceding ones: i.e. the venous artery (*arteria venosa*), which also is badly named, being nothing other than a vein which comes from the lungs where it is divided into many branches, interlaced with those of the arterial vein and those of the tube called the windpipe, through which enters the air we breathe; and the great artery (*aorta*) which, coming out of the heart, sends its branches all through the body. I would also like them to be shown carefully the eleven little valvules which, like so many little doors, open and shut the four openings which are in these two cavities, i.e. three at the entrance of the hollow vein where they are so disposed that they can by no means prevent the blood it contains from flowing into the right-hand ventricle out of the heart, and at the same time prevent it completely from coming out of it; three at the entrance of the arterial vein, which, being

placed the other way round, effectively permit the blood
in this compartment to pass into the lungs, but not the
blood which is in the lungs to return to it; and similarly
two others at the entrance of the venous artery, which
allow the blood from the lungs to flow into the left cavity
of the heart, but which prevent its return; and three at the
entrance of the great artery, which allow the blood to
leave the heart but not to return. And there is no need to
look for any other reason for the number of these valvules, beyond the fact that the opening of the venous artery,
being oval, on account of its position, can conveniently
be closed with two, whereas the others, being round, can
be more conveniently closed by three. In addition, I would
like such persons to have pointed out to them that the
grand artery and the arterial vein are of a much harder and
firmer texture than the venous artery and the hollow vein;
and that the two latter become larger before entering the
heart and that they form there, as it were, two pouches,
called the auricles of the heart, which are made of a substance similar to that of the heart itself; and that there is
always more heat in the heart than anywhere else in the
body; and finally that this heat is capable of causing any
drop of blood which enters the heart's cavities immediately to expand and dilate, in the same way that all liquids
do when they are allowed to fall drop by drop into some
very hot vessel.

For, after that, I do not need to say anything else to
explain the movement of the heart, except that when its
cavities are not full of blood, blood necessarily flows
from the hollow vein into the right and from the venous
artery into the left; because these two vessels are always
full of blood, and their openings which look towards the
heart cannot then be blocked. But as soon as two drops of
blood have thus entered the heart, one into each cavity,
these drops, which cannot be other than very large because

the openings by which they enter are very wide, and the vessels from which they come very full of blood, rarefy and dilate because of the heat they find there: by means of which, making the whole heart expand, they push against and close the five little doors which are at the entrances of the two vessels from which they flowed, thus preventing any more blood coming down into the heart, and, continuing to become more and more rarefied, they push open the six other little doors which are at the entrance to the two other vessels through which they leave the heart, in this way making all the branches of the arterial vein and the great artery swell, almost at the same moment as the heart, which straightaway afterwards contracts, as do these arteries also, because the blood which has entered them has cooled, and their six little doors shut again and the five of the hollow vein and the venous artery open again and let in two more drops of blood, which immediately make the heart and arteries expand, as before. And, because the blood which thus comes into the heart passes through these two pouches called auricles, so it comes about that their movement is contrary to that of the heart, and that they contract when the heart expands. Finally, so that those who do not know the force of mathematical demonstrations and who are not accustomed to distinguishing true reasons from mere verisimilitudes should not venture to deny this without examining it, I would like to warn them that this movement that I have just explained follows as necessarily from the mere disposition of the organs which may be observed in the heart by the eye alone, and from the heat which one can feel there with one's fingers, and from the nature of the blood which can be known by experience, as does that of a clock from the power, the position and the shape of its counterweights and its wheels.

But, if it be asked how it is that the blood in the veins

is not exhausted, flowing in this way continually into the
heart, and how it is that the arteries are not overfilled
since all the blood which passes through the heart goes
into them, I need only reply what an English doctor
has already written about it, to whom must be given
praise for having broken the ice on this topic, and for
being the first who has taught that there are several little
passages at the extremities of the arteries through which
the blood which they receive from the heart enters the
little branches of the veins, whence it returns immediately
to the heart, so that its course is nothing other than a
perpetual circulation. He proves this very well by the
common experience of surgeons who, having bound the
arm moderately tightly above the place at which they open
the vein, cause more blood to come out than if they had
not bound the arm, and the opposite result would occur
if they bound the arm below, between the hand and the
opening, or if they bound it above the opening very
tightly. For it is obvious that a moderate binding, while
being able to prevent the blood which is already in the
arm from returning to the heart through the veins, cannot on that account prevent fresh blood from arriving
from the arteries, because they are situated below the
veins and because their texture, being harder than that
of the veins, is less easy to compress, and also because the
blood which comes from the heart tends to pass through
the arteries to the hand with greater force than it does
when returning from the hand to the heart by the veins.
And since this blood comes out of the arm through the
opening in one of the veins, there must necessarily be
some passages below the ligature, that is to say, towards
the extremities of the arm, through which it can come
from the arteries. He also proves very well what he says
about the circulation of the blood, by certain little pelli-
cles which are so situated in various places along the

veins that they do not allow the blood to pass from the centre of the body towards the extremities, but only to return from the extremities towards the heart; and moreover, from experience which shows that all the blood in the body can come out of it in a very short space of time through one artery if it is cut, even though it were tightly bound very near to the heart and cut between the heart and the ligature, so that one could not possibly have reason to imagine that the blood which flowed out came from anywhere other than the heart.

But there are many other things which bear witness that the true cause of this movement of the blood is as I have said: thus, firstly, the difference to be observed between the blood which comes out of the veins and that which issues from the arteries can only be due to the fact that, being rarefied and, as it were, distilled in passing through the heart, it is thinner, and more vigorous, and warmer straight after leaving the heart, that is to say, while it is in the arteries, than it is just before it goes into the heart, that is to say, in the veins; and, if one looks closely, one will find that this difference appears clearly only near the heart, and not so much so in the more remote parts of the body. In the next place, the hardness in texture of the arterial vein and the great artery shows sufficiently that the blood beats against them with more force than against the veins. And why should the left cavity of the heart and the great artery be larger and wider than the right-hand cavity and the arterial vein, if it were not that the blood from the venous artery, having been only in the lungs since it left the heart, is thinner and is rarefied to a greater degree, and more easily, than the blood which comes direct from the hollow vein? And what can the physicians discover from taking the pulse unless they know that, as the blood changes its nature, it can be rarefied by the heat of the heart more or less strongly and more or less quickly than before? And if one examines how this heat is communicated to the other organs of the body, must one not admit that it is by means of the blood, which, passing through the heart, is reheated and from there spreads throughout the body; from which it arises that if one removes the blood from some part, one removes the heat at the same time; and although the heart were as hot as a piece of glowing iron it would not suffice to warm the feet and hands as it does at present, unless it sent to these extremities a continuous supply of new blood. Then, also, one knows from this that the true use of respiration is to carry enough fresh air to the lungs, to enable the blood which enters the lungs from the right-hand cavity of the heart, where it has been rarefied and, as it were, changed into vapours, to thicken and convert once more into blood before falling into the left-hand cavity, without which it would not be fit to nourish the fire which is there. This is confirmed by the fact that we see that animals which do not have lungs, have only one cavity in the heart, and that children who cannot use their lungs while they are in their mothers' wombs, have an aperture through which blood flows from the hollow vein into the left-hand cavity of the heart, and a pipe through which it travels from the arterial vein into the great artery, without passing through the lung. Then how could digestion take place in the stomach if the heart did not send heat there by the arteries, and with it some of the more fluid parts of the blood, which help to dissolve the foods that have been put there? And is it not easy to understand the action which converts the juice of these foods into blood, if one considers that it is distilled in passing again and again through the heart, perhaps more than a hundred or two hundred times a day? And what else does one need to explain the feeding of the body with blood and the production of the various humours which are in the body other than to say that the force with which the blood, as it is rarefied, passes from the heart towards the extremities of the arteries, causes some of its parts to remain among those of the various members in which they find themselves, and taking the place there of others which they expel; and that according to the position, or the shape, or the smallness of the pores they encounter, some, rather than others, go to certain places, in the same way that, as everyone will have observed, different sieves which, having been pierced with different-sized holes, serve to separate different grains from each other? And, finally, what is most to be noticed in all this is the generation of the animal spirits, which are like a very subtle wind, or rather like a very pure and lively flame which, rising continually in great abundance from the heart to the brain, goes by means of the nerves into the muscles and gives movement to all the limbs, without its being necessary to suppose any other reason why it is that the parts of the blood which, being the most active and most penetrating, are the best suited to compose these spirits, direct themselves to the brain rather than elsewhere, than the fact that the arteries which carry them there are those which come from the heart in the straightest line of all, and that, according to the rules of mechanics, which are the same as the rules of nature, when many things tend together to move towards a same place where there is not enough space for them all, as in the case of the parts of the blood which leave the left-hand cavity of the heart and flow towards the brain, the weakest and least lively must be turned away from it by the strongest, which in this way alone arrive there.

I had explained all these things in some detail in the treatise which I earlier had it in mind to publish. And I had then shown there what must be the structure of the nerves and muscles of the human body, to enable the

animal spirits, contained in the body, to have the power
to move its limbs, as when one sees heads, shortly after
having been cut off, still moving and biting the ground,
even though they are no longer alive; what changes must
take place in the brain to cause waking, sleep and dreams;
how light, sounds, smells, tastes, heat and all the other
qualities of external objects can imprint different ideas
in the brain by means of the senses; how hunger, thirst
and the other internal passions, can also transmit to it
different ideas; what must be taken for the common sense
in which these ideas are received, for memory which
conserves them, and for the imagination, which can
change them in different ways and make up new ones,
and by the same means, distributing the animal spirits
in the muscles, can make the limbs of this body move in
as many different ways, and in a manner as suited to the
objects which are presented to its senses, or to its internal
passions, as our own bodies can move without the will
conducting them. This will not appear in any way strange
to those who, knowing how many different automata or
moving machines the industry of man can devise, using
only a very few pieces, by comparison with the great
multitude of bones, muscles, nerves, arteries, veins and
all the other parts which are in the body of every animal,
will consider this body as a machine, which, having been
made by the hands of God, is incomparably better ordered,
and has in it more admirable movements than any of those
which can be invented by men. And here I gave particular
emphasis to showing that, if there were such machines
which had the organs and appearance of a monkey or of
some other irrational animal, we would have no means of
recognizing that they were not of exactly the same nature
as these animals: instead of which, if there were machines
which had a likeness to our bodies and imitated our
actions, inasmuch as this were morally possible, we would

still have two very certain means of recognizing that they were not, for all that, real men. Of these the first is, that they could never use words or other signs, composing them as we do to declare our thoughts to others. For one can well conceive that a machine may be so made as to emit words, and even that it may emit some in relation to bodily actions which cause a change in its organs, as, for example, if one were to touch it in a particular place, it may ask what one wishes to say to it; if it is touched in another place, it may cry out that it is being hurt, and so on; but not that it may arrange words in various ways to reply to the sense of everything that is said in its presence, in the way that the most unintelligent of men can do. And the second is that, although they might do many things as well as, or perhaps better than, any of us, they would fail, without doubt, in others, whereby one would discover that they did not act through knowledge, but simply through the disposition of their organs: for, whereas reason is a universal instrument which can serve on any kind of occasion, these organs need a particular disposition for each particular action; whence it is that it is morally impossible to have enough different organs in a machine to make it act in all the occurrences of life in the same way as our reason makes us act.

Now by these two same means one can also tell the difference between men and beasts. For it is particularly noteworthy that there are no men so dull-witted and stupid, not even imbeciles, who are incapable of arranging together different words, and of composing discourse by which to make their thoughts understood; and that, on the contrary, there is no other animal, however perfect and whatever excellent dispositions it has at birth, which can do the same. Nor does this arise for lack of organs, for one sees that magpies and parrots can utter words as we do, and yet cannot speak as we do, that is to say, by showing that what they are saying is the expression of thought; whereas men, born deaf and dumb, deprived as much as, or more than, the animals of the organs which in others serve for speech, habitually invent for themselves certain signs, by means of which they make themselves understood by those who, being fairly continuously in their company, have the time to learn their language. And this shows not only that animals have less reason than men, but that they have none at all; for we see that very little of it is required in order to be able to speak; and since one notices inequality among animals of the same species as well as among men, and that some are easier to train than others, it is unbelievable that the most perfect monkey or parrot of its species should not equal in this the most stupid child, or at least a child with a disturbed brain, unless their souls were not of an altogether different nature from our own. And one should not confuse words with the natural movements which bear witness to the passions and can be imitated by machines as well as by animals; neither should one think, as did certain of the Ancients, that animals speak although we do not understand their language. For, if it were so, as they have many organs similar to our own, they could make themselves understood by us as well as by their fellows. It is also particularly noteworthy that although there are many animals which show more skill than we do in certain of their actions, yet the same animals show none at all in many others; so that what they do better than we do does not prove that they have a mind, for it would follow that they would have more reason than any of us and would do better in everything; rather it proves that they do not have a mind, and that it is nature which acts in them according to the disposition of their organs, as one sees that a clock, which is made up of only wheels and springs, can count the hours and measure time more exactly than we can with all our art.

After this I had described the reasonable soul, and shown that it could not in any way be derived from the power of matter, as the other things of which I had spoken, but that it must be created expressly; and I had shown how it is not sufficient that it should be lodged in the human body, like a pilot in his ship, unless perhaps to move its limbs, but that it needs to be joined and united more closely with the body, in order to have, besides, sensations and appetites like our own, and in this way to constitute a true man. Finally, I treated at some length here the subject of the soul, because it is of the greatest importance: for, after the error of those who deny the existence of God, which error I think I have sufficiently refuted above, there is nothing which leads feeble minds more readily astray from the straight path of virtue than to imagine that the soul of animals is of the same nature as our own, and that, consequently, we have nothing to fear or to hope for after this life, any more than have flies or ants; instead, when one knows how much they differ, one can understand much better the reasons which prove that our soul is of a nature entirely independent of the body, and that, consequently, it is not subject to die with it; then, since one cannot see other causes for its destruction, one is naturally led to judge from this that it is immortal.

6

It is now three years since I completed the treatise containing all these things, and I was beginning to revise it so as to be able to put it into the hands of a printer when I learned that persons to whom I defer, and whose authority over my actions is hardly less influential than my own reason over my thoughts, had disapproved of an opinion in a matter of physics, published a little earlier by another person, of which opinion I would not say that I was, but that I had not observed anything in it, before their having censured it, that I could have imagined to be prejudicial either to religion or to the State, or which, consequently, might have stopped me from writing it if reason had persuaded me it was correct; and this made me fear that there might likewise be among my own opinions some one in which I had been mistaken, notwithstanding the great care I have always taken never to accept any new beliefs for which I had not most certain proofs and never to write about any which might be the occasion of disturbance for anyone. This has been sufficient to make me feel obliged in conscience to alter the resolution I had made to publish them. For, although the reasons I had for taking the decision to publish earlier were very strong, my natural inclination, which has always made me hate the business of writing books, made me immediately find sufficient other reasons for not publishing. And these reasons, both for and against, are such, that not only is it to some extent my interest to set them down here, but perhaps also that of the public to know them.

I have never set much store on the things which have come from my own mind, and so long as I reaped no other harvest from the method that I use, apart from satisfying myself about certain difficulties in the field of the speculative sciences, or trying to regulate my conduct according to the precepts which the method taught me, I never considered myself to be under an obligation to write anything about it. For, as far as behaviour is concerned, everyone is so sure that he knows best, that as many reformers as heads might be found, if it were permitted to any to undertake to make changes in it, apart from those whom God had established as rulers over his peoples, or to whom he has given sufficient grace and zeal to be prophets; and although I was very well pleased with my speculations, I believed that others had their own which perhaps pleased them even more. But, as soon as I had acquired some general ideas about physics, and, beginning to test them on various particular difficulties, I observed how far they can lead and how they differ from the principles which have been used up to now, I believed I could not keep them hidden without sinning considerably against the law which obliges us to procure, by as much as is in us, the general good of all men. For they have made me see that it is possible to arrive at knowledge which is most useful in life, and that, instead of the speculative philosophy taught in the Schools, a practical philosophy can be found by which, knowing the power and the effects of fire, water, air, the stars, the heavens and all the other bodies which surround us, as distinctly as we know the various trades of our craftsmen, we might put them in the same way to all the uses for which they are appropriate, and thereby make ourselves, as it were, masters and possessors of nature. Which aim is not only to be desired for the invention of an infinity of devices by which we might enjoy, without any effort, the fruits of the earth and all its commodities, but also principally for the preservation of health, which is undoubtedly the first good, and the foundation of all the other goods of this life; for even the mind depends so much on the temperament and on the disposition of the organs of the body, that if it is possible to find some means of rendering men as a whole wiser and more dexterous than they have been hitherto, I believe it must be sought in medicine. It is true that the medicine practised now contains little of notable use; but without intending to do it any dishonour, I am sure there is no one, even among those who practise it, who does not admit that what is known of it is almost nothing compared to what remains to be known, and that we could free ourselves of an infinity of illnesses, both of the body and of the mind, and even perhaps also of the decline of age, if we knew enough about their causes and about all the remedies with which nature has provided us. So, intending to devote my whole life in the search for so necessary a science, and having come across a path which seems to me such that, by following it, one must inevitably find one's goal, provided one is not prevented either by the shortness of life or by the lack of experiments, I judged that there was no better remedy against these two obstacles than faithfully to communicate to the public all the little I had found, and to urge good minds to try to go beyond this in contributing, each according to his inclination and his capacity, to the experiments which must be made, and communicating also to the public everything they learned; so that, the last beginning where their predecessors had left off, and thereby linking the lives and the labours of many, we might all together go much further than each man could individually.

Moreover I noticed, concerning experiments, that they are all the more necessary the more one is advanced in knowledge. For, to begin with, it is better to use only what presents itself spontaneously to our senses and of which we cannot remain ignorant provided we give it even a moment's reflection, than to seek out more rare and more abstruse phenomena; the reason for this is that these rarer ones are often misleading when one does not yet know the causes of the commoner ones, and that the circumstances on which they depend are almost always so special and so minute that they are very difficult to detect. But the order that I have adhered to in this has been as follows: firstly, I have tried to find in general the principles or first causes of everything which is, or which may be, in the world, without considering to this end anything but God alone, who has created it, or taking them from any other source than from certain seeds of truths which are naturally in our minds. After that, I examined what were the first and most ordinary effects that could be deduced from these causes, and it seems to me that in this way I have found heavens, stars, an earth, and even on the earth water, air, fire, minerals and other similar things, which are the most common and simplest of all, and consequently the easiest to know. But, when I wanted to come down to those which were more particular, so many different ones presented themselves to me, that I did not believe it possible for the human mind to distinguish the forms or species of bodies which are on the earth, from an infinity of others which might have been there, if it had been the will of God to put them there, or, consequently, to apply them to our use, unless we reach for causes through effects and make use of many particular experiments. Following which, turning over in my mind all the objects which had ever been presented to my senses, I dare to state that I observed nothing that I could not easily enough explain by means of the principles I had found. But I must also admit that the power of nature is so ample and so vast, and that these principles are so simple and so general, that I observe almost no individual effect without immediately knowing that it can be deduced in many different ways, and that my greatest difficulty is ordinarily to find in which of these ways the effect depends upon them; for to this end I know no other expedient but then to seek certain experiments which are such that their result will not be the same if it is in one of these ways that the explanation lies as if it lies in anóther. Moreover, I have reached the point at which I see well enough, it seems to me, how to go about making most of those experiments which serve this purpose; but I see also that they are such and in so great a number, that neither my hands nor my income, even if I were to have a thousand times more than I have, would suffice for all of them; so that, according to the means I shall henceforth have of doing more or less of them, I shall advance to a greater or lesser degree in knowledge of nature. This is what I promised myself to make known by the treatise which I had written, and to show in it so clearly the use the public could derive from it, that I would oblige all those who wish for the general well-being of men, that is to say, all those who are truly virtuous and not simply in appearance or who merely profess to be so, both to communicate to me the experiments they have already made and to help me to investigate those which remain to be done.

But since then I have had other reasons to change my mind and to think that I ought indeed to continue to record everything I judged to be of some importance, as I discovered the truth about them, and to bring to the task the same care as if I wished to publish. This as much in order to have the greater opportunity of examining them thoroughly, for there is no doubt that one always looks more closely at what one thinks will be seen by many people than at what one does only for oneself – and often the things which have seemed true when I first conceived them have appeared false when I committed them to paper – as in order to lose no opportunity of being useful to the public, if I can, so that, if my writings are worth anything, those who have them after my death may use them as they think most appropriate. But I decided that I could in no way consent to their publication in my lifetime, so that neither the oppositions nor the controversies of which they might be the object, nor even whatever reputation they might gain for me, would give me any occasion to waste the time I intend to use in acquiring knowledge. For, although it is true that every man is obliged to promote the good of others, as far as it is in him to do so, and that to be no use to anyone is really to be worthless, it is none the less true also that our solicitude ought to extend beyond the present time, and that it is good to omit doing things which might perhaps bring some profit to those who are living, when one aims to do other things which will be of greater benefit to posterity. And, in truth, I will not hide the fact that the little I have learned so far is almost nothing compared to what I do not know, and to what I do not despair of being able to learn; for it is almost the same with those who discover truth in the sciences little by little, as with those who, becoming rich, have less trouble in making great acquisitions, than they had earlier, when poorer, in making much smaller ones. Or they can be compared to leaders of armies, whose forces usually grow in proportion to their victories, and who need more skill to maintain themselves after losing a battle than they need, after a victory, to take towns and provinces. For he indeed fights battles who tries to overcome all the difficulties and errors which prevent him from arriving at the knowledge of truth, and he loses a battle when he accepts some false opinion in a subject which is of some generality and importance. It takes much more skill afterwards to recover one's former position than it takes to make great progress when one is already in possession of principles which are assured. For myself, if I have already found any truths in the sciences (and I hope the things contained in this volume will prove that I have found a few), I can say that they are only the results and consequences of five or six principal difficulties which I have overcome, and that I count as so many battles in which I had victory on my side. I will not even fear to say that I believe I need only to win two or three more like them in order to reach completely my goal, and that I am not so advanced in age that, according to the ordinary course of nature, I may not still have enough leisure to this end. But I consider myself all the more obliged to organize prudently the time which remains to me, the greater hope I have of being able to use it well, and no doubt I should have many opportunities of wasting time if I were to publish the principles of my physics. For although they are almost all so evident that one has only to understand them to accept them, and although there is not one among them which I do not think I could demonstrate, yet, as it is impossible that they should be in accord with all the various opinions of others, I foresee that I would frequently be distracted by the oppositions they would engender.

One may say that these oppositions would be useful both in making me aware of my errors, and, if I had something sound to say, in better acquainting others with it; and as many men can see more than one man, by beginning henceforth to use my principles, they might help me in turn with their own discoveries. But although I recognize my extreme liability to error, and almost never trust the first thoughts which come to me, nevertheless the experience I have had of objections which may be made to my views, prevents me from expecting any profit from them. For I have already often experienced the judgements, both of those I have taken to be my friends and of some others to whom I thought I was an object of indifference, and even also of some whose malice and envy would make them try to reveal what partiality hid from my friends; but it has rarely happened that any objection has been put forward that I had not altogether foreseen, unless it were far removed from my subject, so that I have hardly ever encountered any critic of my opinions who did not seem either less exacting or less equitable than myself. And, moreover, I have never noticed that, by the method of disputation practised in the Schools, any truth has been discovered of which one was ignorant before. For, while each strives after victory, he is much more preoccupied in making the best of verisimilitudes than in weighing the reasons on both sides; and those who have been good advocates for a long time are not, for all that, subsequently better judges.

As for the usefulness to others of the communication of my thoughts, it could not be very great, inasmuch as I have not yet taken them so far that much does not need to be added to them before they could be applied in practice. And I think I can say without vanity that, if there is anyone capable of bringing them to this point, it must be myself rather than any other; not that there may not be in the world many minds incomparably better than mine, but because one cannot so well grasp a thing and make it one's own when it is learnt from another person, as when one discovers it oneself. This is so true of this matter that, although I have often explained some of my opinions to people of good mind, and who, while I was speaking to them, seemed to understand most distinctly, yet, when they repeated these opinions, I have noticed that they almost always change them in such a way that I could no longer acknowledge them as mine; I am glad to take this opportunity to ask future generations never to believe that the things people tell them come from me, unless I myself have published them; and I am not in the least astonished at the extravagances attributed to all those ancient philosophers whose writings we do not have, neither do I judge on that account that their thoughts were extremely unreasonable, seeing that they were the best brains of their time, but only that they have been misrepresented to us. For one sees also that it has almost never happened that any one of their disciples has surpassed them; and I am convinced that the most devoted of those who now follow Aristotle would think themselves happy if they had as much knowledge of nature as he had, even if it were a condition that they would never have more. They are like the ivy which does not seek to climb higher than the trees which support it, and which even often comes down again after reaching the top; for it seems to me that those people come down again, that is to say, become in some way less learned than if they abstained from study, who, not content with knowing all that is intelligibly explained in their author, wish, in addition, to find in him the solution of many difficulties of which he says nothing and about which he perhaps never thought. However, their fashion of philosophizing is most convenient for those who have only very mediocre minds; for the obscurity of the distinctions and principles which they use enables them to speak about all things as boldly as if they really knew them, and to maintain everything they say against the subtlest and most skilful, without anyone being able to convince them of their error. In this they seem to me like a blind man who, in order to fight a person who can see, without disadvantage, brings him into the depths of a very dark cellar; and I may say that these people have an interest in my abstention from publishing the principles of the philosophy I use, for, as these principles are very simple and evident, I would be doing almost the same, in publishing them, as if I opened a few windows and let the light of day into this cellar into which they have descended in order to fight. But even the best minds have no reason to wish to know these principles, for, if they wish to be able to talk about all things, and to acquire the reputation of being learned, they will achieve this more easily by contenting themselves with verisimilitudes, which can be found without much trouble in all kinds of matters, than by seeking the truth, which is revealed only little by little and in a few matters, and which, when other matters arise, obliges one to confess frankly one's ignorance. If, however, they prefer knowledge of a handful of truths to the vanity of appearing to be ignorant of none, as is undoubtedly preferable, and if they agree to follow a course similar to mine, they do not need me to say anything more, for this purpose, than I have already said in this discourse. For, if they are capable of going further than I have, they will be capable also, *a fortiori*, of discovering for themselves all that I think I have found; especially as, never having examined anything except in order, it is certain that what remains for me to discover is in itself more difficult and more remote than what I have encountered up to now, and it would give them much less pleasure to learn it from me than for themselves. Besides, the habit they will acquire in seeking first easy things, and passing little by little by degrees to other more difficult things, will be of more use to them than all my instructions could be. As for me, I am persuaded that if I had been taught from youth all the truths of which I have since sought demonstrations, and had had no trouble in learning them, I might never have known any others; at least I should never have acquired the habit and facility which I think I have of ever discovering new truths, according as I apply myself to looking for them. And, in a word, if there is any task in the world which cannot be completed by any other person as well as it can by the person who began it, it is that at which I am working.

It is true that, concerning the experiments which may serve this purpose, one man alone could not suffice to perform them all; but equally he could not usefully employ other hands than his own, unless those of artisans or such people as he could pay, and whom the hope of gain, which is a very effective means, would cause to perform accurately all the things which he laid down for them. For, as for the volunteers who would perhaps offer to help him, out of curiosity or the wish to learn, apart from being usually more full of promises than they are effective, and having only fine ideas of which none ever comes to anything, they would inevitably wish to be recompensed by the explanation of certain difficulties, or at least by compliments and useless conversations which could not cost him so little of his time that he would not lose by it. And as for the experiments already made by others, even if these people wish to communicate them to him, which those who consider them as secrets would never do, they are for the most part made up of so many circumstances and superfluous considerations that it would cause him no small difficulty to disentangle what truth they contain; besides, he would find almost all of them so badly explained or even so false, because those who made them have attempted to make them appear in conformity with their principles, that, even if there were a few which he could make use of, they could not be worth the time he would have to employ in selecting them. So that, if there were someone in the world whom one knew certainly to be capable of discovering the greatest, and most useful discoveries, and if therefore other men made every effort to help him achieve his designs, I do not see that they could do anything for him except to contribute to the cost of the experiments he would need to do, and for the rest, to prevent his leisure from being taken away from him by the importunity of anyone. But, while not being so presumptuous as to be willing to promise anything extraordinary, or feeding myself on such vain fancies as to imagine that the public must be much interested in my plans, I do not, on the other hand, have so base a soul as to wish to accept from anyone whatsoever any favour that I might not be deemed to have merited.

All these considerations, added together, were the reason why, three years ago, I did not wish to publish the treatise which I had on hand, and why I even resolved not to reveal any other during my lifetime, which was so general, or from which the foundations of my physics might be understood. But there have been since then two other reasons which have compelled me to set down here some particular specimens and to give the public some account of my actions and my plans. The first of these is that, if I omitted to do so, many people who have known of my earlier intention to publish some writings might imagine that the reasons why I have not done so are more dishonourable to me than they are. For, although I have no excessive love of fame, or even, if I dare say so, though I hate it, in so far as I judge it to be contrary to repose which I value above everything, yet, at the same time, I have never tried to conceal my actions as if they were crimes, neither have I taken much precaution to be unknown; and this as much because I should have thought I would be doing myself an injustice, as because it would have given me some sort of disquiet, which would again have been contrary to the perfect peace of mind which I seek. And since, while being thus indifferent to being known or not, I have been unable to avoid acquiring some sort of reputation, I thought I ought to do my best at least to save myself from having a bad one. The other reason which has compelled me to put pen to paper is that, becoming daily more and more conscious of the delay which my plan of self-instruction is suffering, owing to the vast number of experiments which I need to make, and which it is impossible for me to do without help from others, although I do not flatter myself so much as to hope that the public will take a great share in my interests, I am yet unwilling to default so much in my own cause, as to give those who will survive me reason to reproach me one day for not having left them many things in a much better state than I have done, if I had not too much neglected to acquaint them of the ways in which they could have contributed to my designs.

And I thought that it was easy for me to choose some matters which, without being too much open to controversy, or obliging me to make known more of my principles than I wished, would yet show clearly enough what I can or cannot do in the sciences. I cannot say whether I have succeeded in this or not, and I do not wish to prejudice anyone's judgement by speaking myself of my writings; but I would be very pleased if they were to be examined, and, so that people may feel all the more free to do so, I beg all those who have any objections to make to them, to take the trouble to send them to my publisher, and, on being advised of them by him, I shall try to publish at the same time both the objection and my reply; and, by this means, readers, seeing both together, may judge the truth all the more easily. For I do not promise ever to give long answers, but only to admit my mistakes most frankly, if I perceive them, or, if I cannot, to say simply what I believe to be necessary for the defence of what I have written, without introducing the explanation of any new matter so as to avoid engaging myself in endless discussion from one topic to another.

And if some of the matters of which I have spoken at the beginning of the *Dioptrics* and *Meteorics* shock the reader at first sight, because I call them hypotheses, and do not seem to want to prove them, I request him to read the whole patiently and attentively, and I hope that he will be satisfied; for it seems to me that my reasonings follow each other in such a way that, as the last are demonstrated by the first, which are their causes, the first are proved, reciprocally, by the last, which are their effects. And one must not imagine that in this matter I commit the fallacy which logicians call a circle, for, experience rendering most of these effects very certain, the causes from which I deduce them do not serve so much to prove as to explain them; on the contrary, it is the reality of the causes which is proved by the reality of the effects. And I have called them hypotheses only so that it may be known that, while I think I can deduce them from these first truths which I have explained above, I have expressly decided not to do so, in order to prevent certain minds who imagine that they know in a day all that it has taken another person twenty years to think out, as soon as he has told them two or three words about it, and who are all the more liable to err, and the less capable of the truth the more penetrating and lively they are, from taking the opportunity to build some extravagant philosophy on what they believe to be my principles, and to avoid my being blamed for it. For, as to the opinions which are entirely mine, I offer no apology for their being new, since, if their reasons be properly examined, I am convinced that they will be found so simple and so in conformity with common sense that they will seem less extraordinary and less strange than any others one might have on the same subjects. Neither do I boast of being the first discoverer of any of them, but only of having accepted them, neither because they had been expressed by others, nor because they had not been, but only because reason convinced me of their truth. And if artisans are unable immediately to execute the invention which is explained in the *Dioptrics*, I do not believe one can say on that account that it is bad; for, inasmuch as skill and practice are needed to make and to adjust the machines that I have described, so that no detail is overlooked, I would be no less astonished if they succeeded at the first attempt than if someone were to learn in one day to play the lute with accomplishment merely because he had been given a good score. And if I write in French, which is the language of my country, rather than in Latin, which is that of my teachers, it is because I hope that those who use only their pure natural reason will be better judges of my opinions than those who believe only in the books of the ancients; and, as for those who unite good sense with study, whom alone I wish to have for my judges, they will not, I feel sure, be so partial to Latin that they will refuse to hear my reasons because I express them in the vulgar tongue.

In conclusion, I do not wish to speak here in detail of the progress I hope to make in the sciences in the future nor to make any promise to the public that I am not certain of being able to fulfil; but I will say simply that I have resolved to devote the time left to me to live to no other occupation than that of trying to acquire some knowledge of Nature, which may be such as to enable us to deduce from it rules in medicine which are more assured than those we have had up to now; and that my inclination turns me away so strongly from all other sorts of projects, and particularly from those which can only be useful to some while being harmful to others, that if any situation arose in which I was forced to engage in such matters, I do not think that I would be able to succeed. On this, I here make a public declaration which I know very well cannot serve to make me of consequence in the world, but then I have no wish to be so; and I shall always hold myself more obliged to those by whose favour I enjoy my leisure unhindered, than to those who might offer me the highest dignities on earth.

MEDITATIONS ON THE FIRST PHILOSOPHY IN WHICH THE EXISTENCE OF GOD AND THE REAL DISTINCTION BETWEEN THE SOUL AND THE BODY OF MAN ARE DEMONSTRATED

FIRST MEDITATION

About the Things We May Doubt

IT is some time ago now since I perceived that, from my earliest years, I had accepted many false opinions as being true, and that what I had since based on such insecure principles could only be most doubtful and uncertain; so that I had to undertake seriously once in my life to rid myself of all the opinions I had adopted up to then, and to begin afresh from the foundations, if I wished to establish something firm and constant in the sciences. But as this undertaking seemed to me very great, I waited until I had attained an age sufficiently mature that I could not hope, at a later stage in life, to be more fit to execute my plan; and this has made me delay so long that I should henceforth consider that I was committing a fault if I were still to use in deliberation the time which remains to me for action.

Now therefore, that my mind is free from all cares, and that I have obtained for myself assured leisure in peaceful solitude, I shall apply myself seriously and freely to the general destruction of all my former opinions. Now it will not be necessary, in order to accomplish this aim, to prove that they are all false, a point which perhaps I would never reach; but inasmuch as reason persuades me already that I must avoid believing things which are not entirely certain and indubitable, no less carefully than those things which seem manifestly false, the slightest ground for doubt that I find in any, will suffice for me to reject all of them. And to this end there will be no need for me to examine each one individually, which would be an endless task; but because the destruction of the foundations necessarily brings down with it the rest of the edifice, I

shall make an assault first on the principles on which all
my former opinions were based.

Everything I have accepted up to now as being absolutely true and assured, I have learned from or through
the senses. But I have sometimes found that these senses
played me false, and it is prudent never to trust entirely
those who have once deceived us.

But, although the senses sometimes deceive us, concerning things which are barely perceptible or at a great
distance, there are perhaps many other things about which
one cannot reasonably doubt, although we know them
through the medium of the senses, for example, that I am
here, sitting by the fire, wearing a dressing-gown, with
this paper in my hands, and other things of this nature.
And how could I deny that these hands and this body
belong to me, unless perhaps I were to assimilate myself
to those insane persons whose minds are so troubled and
clouded by the black vapours of the bile that they constantly assert that they are kings, when they are very poor;
that they are wearing gold and purple, when they are
quite naked; or who imagine that they are pitchers or that
they have a body of glass. But these are madmen, and I
would not be less extravagant if I were to follow their
example.

However, I must here consider that I am a man, and
consequently that I am in the habit of sleeping and of
representing to myself in my dreams those same things,
or sometimes even less likely things, which insane people
do when they are awake. How many times have I dreamt
at night that I was in this place, dressed, by the fire, although I was quite naked in my bed? It certainly seems to
me at the moment that I am not looking at this paper with
my eyes closed; that this head that I shake is not asleep;
that I hold out this hand intentionally and deliberately,
and that I am aware of it. What happens in sleep does not

seem as clear and distinct as all this. But in thinking about it carefully, I recall having often been deceived in sleep by similar illusions, and, reflecting on this circumstance more closely, I see so clearly that there are no conclusive signs by means of which one can distinguish clearly between being awake and being asleep, that I am quite astonished by it; and my astonishment is such that it is almost capable of persuading me that I am asleep now.

Let us suppose, then, that we are now asleep, and that all these particulars, namely, that we open our eyes, move our heads, hold out our hands, and such like actions, are only false illusions; and let us think that perhaps our hands and all our body are not as we see them. Nevertheless, we must at least admit that the things which appear to us in sleep are, as it were, pictures and paintings which can only be formed in the likeness of something real and true; and that therefore these general things at least, namely, eyes, head, hands and all the rest of the body are not imaginary things but are real and existent. For indeed painters, even when they study with the utmost skill to represent Sirens and Satyrs by strange and extraordinary shapes, cannot attribute to them entirely new forms and natures, but only make a certain mixture and compound of the limbs of various animals; or if perhaps their imagination is extravagant enough to invent something so new that we have never seen the like of it, and that, in this way, their work presents us with something purely fictitious and absolutely false, at least the colours of which they have composed it are real. And by the same reasoning, although these general things, viz. eyes, head, hands and the like, may be imaginary, we have to admit that there are even simpler and more universal things which are true and exist, from the mixture of which, no more or less than from the mixture of certain real colours, all the images of things, whether true and real or fictitious and fantastic, which dwell in our thoughts, are formed. Corporeal nature in general, and its extension, are of this class of things: together with the figure of extended things, their quantity or size, and their number, as also the place where they are, the time during which they exist, and such like.

This is why perhaps that, from this, we shall not be wrong in concluding that physics, astronomy, medicine, and all the other sciences which depend on the consideration of composite things, are most doubtful and uncertain, but that arithmetic, geometry and the other sciences of this nature, which deal only with very simple and general things, without bothering about their existence or non-existence, contain something certain and indubitable. For whether I am awake or sleeping, two and three added together always make five, and a square never has more than four sides; and it does not seem possible that truths so apparent can be suspected of any falsity or uncertainty.

Nevertheless, I have for a long time had in my mind the belief that there is a God who is all-powerful and by whom I was created and made as I am. And who can give me the assurance that this God has not arranged that there should be no earth, no heaven, no extended body, no figure, no magnitude, or place, and that nevertheless I should have the perception of all these things, and the persuasion that they do not exist other than as I see them? And, further, as I sometimes think that others are mistaken, even in the things they think they know most certainly, it is possible that God has wished that I should be deceived every time I add two and three or count the sides of a square, or form some judgement even simpler, if anything simpler than that can be imagined. But perhaps God has not wished me to be deceived in this way, for he is said to be supremely good. However, if it were in contradiction to his goodness to have made me in such a way that I always deceived myself, it would seem also to be contrary to his goodness to allow me to be wrong sometimes, and nevertheless it is beyond doubt that he permits it.

There will be some perhaps who would prefer to deny the existence of a God so powerful than to believe that all other things are uncertain. But let us not oppose them for the moment, and let us suppose in their favour that everything said here about a God is a fable. Nevertheless, however they suppose that I reached the state and being which I possess, whether they attribute it to some destiny or fate, or to chance or to a continuous sequence and conjunction of events, it is certain that, because fallibility and error are a kind of imperfection, the less powerful the author to whom they attribute my origin, the more probable it will be that I am so imperfect as to be deceived all the time. I have certainly nothing to say in reply to such reasonings, but am constrained to avow that, of all the opinions that I once accepted as true, there is not one which is not now legitimately open to doubt, not through any lack of reflection or lightness of judgement, but for very strong and deeply considered reasons; so that if I wish to find anything certain and assured in the sciences, I must from now on check and suspend judgement on these opinions and refrain from giving them more credence than I would do to things which appeared to me manifestly false.

But it is not enough to have made these observations; I must also take care to remember them; for those old and customary opinions still recur often in my mind, long and familiar usage giving them the right to occupy my mind against my will and, as it were, to dominate my mind. And I shall never rid myself of the habit of acquiescing in them and of having confidence in them so long as I look upon them as what in fact they are, that is to say, in some degree doubtful, as I have just shown, and yet highly probable, so that it is more reasonable to believe than to deny them. This is why I think I shall proceed more prudently if, taking an opposite course, I endeavour to deceive myself, pretending that all these opinions are false and imaginary, until, having so balanced my prejudices that they may not make my judgement incline more to one side than to another, my judgement may no longer be overpowered as hitherto by bad usage and turned from the right path which can lead it to the knowledge of truth. For I am assured that, meanwhile, there can be no danger or error in this course, and that, for the present, it would be impossible to press my distrust too far, for it is not now action I seek as my end but simply meditation and knowledge.

I shall suppose, therefore, that there is, not a true God, who is the sovereign source of truth, but some evil demon, no less cunning and deceiving than powerful, who has used all his artifice to deceive me. I will suppose that the heavens, the air, the earth, colours, shapes, sounds and all external things that we see, are only illusions and deceptions which he uses to take me in. I will consider myself as having no hands, eyes, flesh, blood or senses, but as believing wrongly that I have all these things. I shall cling obstinately to this notion; and if, by this means, it is not in my power to arrive at the knowledge of any truth, at the very least it is in my power to suspend my judgement. This is why I shall take great care not to accept into my belief anything false, and shall so well prepare my mind against all the tricks of this great deceiver that, however powerful and cunning he may be, he will never be able to impose on me.

But this undertaking is arduous, and a certain indolence leads me back imperceptibly into the ordinary course of life. And just as a slave who was enjoying in his sleep an imaginary freedom, fears to be awakened when he begins to suspect that his liberty is only a dream, and conspires with these pleasant illusions to be deceived by them longer, so I fall back of my own accord into my former opinions, and fear to awake from this slumber lest the laborious wakeful hours which would follow this peaceful rest, instead of bringing me any light of day into the knowledge of truth, would not be sufficient to disperse the shadows caused by the difficulties which have just been raised.

SECOND MEDITATION

*Of the Nature of the Human Mind; and that it
is Easier to Know than the Body*

THE Meditation of yesterday has filled my mind with so
many doubts that it is no longer in my power to forget
them. And yet I do not see how I shall be able to resolve
them; and, as though I had suddenly fallen into very deep
water, I am so taken unawares that I can neither put my
feet firmly down on the bottom nor swim to keep myself
on the surface. I make an effort, nevertheless, and follow
afresh the same path upon which I entered yesterday, in
keeping away from everything of which I can conceive
the slightest doubt, just as if I knew that it was absolutely
false; and I shall continue always in this path until I have
encountered something which is certain, or at least, if I
can do nothing else, until I have learned with certainty
that there is nothing certain in the world.

Archimedes, in order to take the terrestrial globe from
its place and move it to another, asked only for a point
which was fixed and assured. So also, I shall have the
right to entertain high hopes, if I am fortunate enough
to find only one thing which is certain and indubitable.

I suppose therefore that all the things I see are false; I
persuade myself that none of those things ever existed
that my deceptive memory represents to me; I suppose I
have no senses; I believe that body, figure, extension,
movement and place are only fictions of my mind. What,
then, shall be considered true? Perhaps only this, that
there is nothing certain in the world.

But how do I know there is not some other thing,
different from those I have just judged to be uncertain,

about which one could not have the slightest doubt? Is there not a God, or some other power, which puts these thoughts into my mind? But that is unnecessary, for perhaps I am capable of producing them myself. Myself, then, at least am I not something? But I have already denied that I have any senses or any body. I hesitate, however, for what follows from that? Am I so dependent on body and senses that I cannot exist without them? But I had persuaded myself that there was nothing at all in the world: no sky, no earth, no minds or bodies; was I not, therefore, also persuaded that I did not exist? No indeed; I existed without doubt, by the fact that I was persuaded, or indeed by the mere fact that I thought at all. But there is some deceiver both very powerful and very cunning, who constantly uses all his wiles to deceive me. There is therefore no doubt that I exist, if he deceives me; and let him deceive me as much as he likes, he can never cause me to be nothing, so long as I think I am something. So that, after having thought carefully about it, and having scrupulously examined everything, one must then, in conclusion, take as assured that the proposition: *I am, I exist*, is necessarily true, every time I express it or conceive of it in my mind.

But I, who am certain that I am, do not yet know clearly enough what I am; so that henceforth I must take great care not imprudently to take some other object for myself, and thus avoid going astray in this knowledge which I maintain to be more certain and evident than all I have had hitherto.

For this reason, I shall now consider afresh what I thought I was before I entered into these last thoughts; and I shall retrench from my former opinions everything that can be invalidated by the reasons I have already put forward, so that absolutely nothing remains except that which is entirely indubitable. What, then, did I formerly think I was? I thought I was a man. But what is a man? Shall I say rational animal? No indeed: for it would be necessary next to inquire what is meant by animal, and what by rational, and, in this way, from one single question, we would fall unwittingly into an infinite number of others, more difficult and awkward than the first, and I would not wish to waste the little time and leisure remaining to me by using it to unravel subtleties of this kind. But I shall rather stop to consider here the thoughts which sprang up hitherto spontaneously in my mind, and which were inspired by my own nature alone, when I applied myself to the consideration of my being. I considered myself, firstly, as having a face, hands, arms, and the whole machine made up of flesh and bones, such as it appears in a corpse and which I designated by the name of body. I thought, furthermore, that I ate, walked, had feelings and thought, and I referred all these actions to the soul; but I did not stop to consider what this soul was, or at least, if I did, I imagined it was something extremely rare and subtle, like a wind, flame or vapour, which permeated and spread through my most substantial parts. As far as the body was concerned, I was in no doubt as to its nature, for I thought I knew it quite distinctly, and, if I had wished to explain it according to the notions I had of it, I would have described it in this way: by body, I understand all that can be terminated by some figure; that can be contained in some place and fill a space in such a way that any other body is excluded from it; that can be perceived, either by touch, sight, hearing, taste or smell; that can be moved in many ways, not of itself, but by something foreign to it by which it is touched and from which it receives the impulse. For as to having in itself the power to move, to feel and to think, I did not believe in any way that these advantages might be attributed to corporeal nature; on the contrary, I was somewhat astonished to see that such faculties were to be found in certain bodies.

But as to myself, who am I, now that I suppose there is someone who is extremely powerful and, if I may so say, malicious and cunning, who employs all his efforts and industry to deceive me? Can I be sure of having the least of all the characteristics that I have attributed above to the nature of bodies? I pause to think about it carefully, I turn over all these things in my mind, and I cannot find one of which I can say that it is in me. There is no need for me to stop and enumerate them. Let us pass, then, to the attributes of the soul, and see if there are any of these in me. The first are eating and walking; but if it is true that I have no body, it is true also that I cannot walk or eat. Sensing is another attribute, but again this is impossible without the body; besides, I have frequently believed that I perceived in my sleep many things which I observed, on awakening, I had not in reality perceived. Another attribute is thinking, and I here discover an attribute which does belong to me; this alone cannot be detached from me. *I am, I exist*: this is certain; but for how long? For as long as I think, for it might perhaps happen, if I ceased to think, that I would at the same time cease to be or to exist. I now admit nothing which is not necessarily true: I am therefore, precisely speaking, only a thing which thinks, that is to say, a mind, understanding, or reason, terms whose significance was hitherto unknown to me. I am, however, a real thing, and really existing; but what thing? I have already said it: a thing which thinks. And what else? I will stir up my imagination in order to discover if I am not something more. I am not this assemblage of limbs called the human body; I am not a thin and penetrating air spread through all these members; I am not a wind, a breath of air, a vapour, or anything at all that I can invent or imagine, since I have supposed that all those things were nothing, and yet, without changing this supposition, I find I am nevertheless certain that I am something.

But also, it may be that these same things that I suppose do not exist, because they are unknown to me, are not in truth different from me whom I know. I do not know; I am not debating this point now. I can judge only of things which are known to me: I have recognized that I exist, and I, who recognize I exist, seek to discover what I am. It is most certain, however, that this notion and knowledge of myself, thus precisely taken, do not depend on things the existence of which is not yet known to me; neither, consequently, and *a fortiori*, do they depend on any of those which are feigned and invented by the imagination. And even these terms feigning and imagining, warn me of my error; for I should be feigning, in truth, if I were to imagine that I am anything, since imagining is nothing other than contemplating the figure or image of a corporeal object. Now I know already for certain that I exist, and at the same time that it is possible that all those images, and, in general, all the things one relates to the nature of body, are nothing but dreams or chimera. From this I see clearly that it is as unreasonable for me to say: I shall stir up my imagination in order to know more distinctly what I am, as to say: I am now awake, and I perceive something real and true; but, because I do not perceive it clearly enough, I shall go to sleep expressly so that my dreams may show this object to me with greater truth and clearness. And in this way, I recognize certainly that nothing of all that I can understand by means of imagination belongs to this knowledge that I have of myself, and that it is necessary to call one's mind back and turn it away from this mode of thinking, so that it can itself recognize its own nature very distinctly.

But what, then, am I? A thing that thinks. What is a

thing that thinks? That is to say, a thing that doubts, perceives, affirms, denies, wills, does not will, that imagines also, and which feels. Indeed this is not a little, if all these properties belong to my nature. But why should they not so belong? Am I not still this same being who doubts of almost everything; who nevertheless understands and conceives certain things; who affirms those alone to be true; who denies all the rest; who wishes and desires to know more of them and does not wish to be deceived; who imagines many things, even sometimes in spite of himself; and who also perceives many, as if through the intermediary of the organs of the body? Is there nothing in all this which is as true as it is certain that I am, and that I exist, even though I were always to be sleeping, and though he who has given me my being should use all his power to deceive me? Is there also any one of these attributes which may be distinguished from my thought, or that one could say was separate from me? For it is so self evident that it is I who doubt, who understand and who wish, that there is no need here to add anything to explain it. And I have equally certainly the power to imagine, for even though it may be (as I have supposed above) that the things I imagine are not true, nevertheless, this capacity for imagining does not cease to be really in me, and forms part of my thinking. Finally, I am the same being who senses, that is to say who apprehends and knows things, as by the sense-organs, since, in truth, I see light, hear noise and feel heat. But it will be said that these appearances are false and that I am dreaming. Let it be so; all the same, at least, it is very certain that it seems to me that I see light, hear a noise and feel heat; and this is properly what in me is called perceiving and this, taken in this precise sense, is nothing other than thinking. From this I begin to know what I am, a little more clearly and distinctly than hitherto.

But I cannot help believing that corporeal objects, whose images are formed by my thoughts, and which come under the senses, are more distinctly known to me than that, I know not what, part of me which does not fall within the grasp of the imagination; although in truth it may seem very strange that things I find doubtful and distant, are more clearly and easily known to me than those which are true and certain, and which belong to my own nature. But I see very well how it is: my mind likes to wander, and cannot yet contain itself within the precise limits of truth. Let us therefore give it its head once more, so that, later on, tightening the rein gently and opportunely, we shall the more easily be able to govern and control it.

Let us begin by considering the most common things, those which we believe we understand the most distinctly, namely, the bodies we touch and see. I am not speaking of bodies in general, for these general notions are usually more confused, but of one body in particular. Let us take, for example, this piece of wax which has just been taken from the hive; it has not yet lost the sweetness of the honey it contained; it still retains something of the smell of the flowers from which it was gathered; its colour, shape and size, are apparent; it is hard, cold, it is tangible; and if you tap it, it will emit a sound. So, all the things by which a body can be known distinctly are to be found together in this one.

But, as I am speaking, it is placed near a flame: what remained of its taste is dispelled, the smell disappears, its colour changes, it loses its shape, it grows bigger, becomes liquid, warms up, one can hardly touch it, and although one taps it, it will no longer make any sound. Does the same wax remain after this change? One must admit that it does remain, and no one can deny it. What, then, was it that I knew in this piece of wax with such distinctness?

Certainly it could be nothing of all the things which I perceived by means of the senses, for everything which fell under taste, smell, sight, touch or hearing, is changed, and yet the same wax remains. Perhaps it was what I now think, namely, that the wax was not the sweetness of honey, or the pleasant smell of flowers, the whiteness, or the shape, nor the sound, but only a body which a little earlier appeared to me in these forms, and which is now to be perceived in other forms. But to speak precisely, what is it that I imagine when I conceive it in this way? Let us consider it attentively, and setting aside everything that does not belong to the wax, let us see what remains. Indeed nothing remains, except something extended, flexible and malleable. Now, what does that mean: flexible and malleable? Is it not that I imagine that this wax, being round, is capable of becoming square, and of passing from a square to a triangular figure? No indeed, it is not that, for I conceive of it as capable of undergoing an infinity of similar changes, and as I could not embrace this infinity by my imagination, consequently this conception I have of the wax is not the product of the faculty of imagination.

What, now, is this extension? Is it not also unknown, since it increases as the wax melts, is greater when the wax is completely melted, and very much greater still when the heat is intensified; and I should not conceive clearly and according to truth what the wax is, if I did not remember that it is capable of taking on more variations in extension that I have ever imagined. I must therefore agree that I could not even conceive by means of the imagination what this wax is, and that it is my understanding alone which conceives it. I say this piece of wax in particular, for, as to wax in general, this is still more evident. Now, what is this wax, which cannot be conceived except by the understanding or mind? Indeed it is the same which

I see, touch, imagine, and which I knew from the start. But, and this is to be noted, the perception of it, or the action by which one perceives it, is not an act of sight, or touch, or of imagination, and has never been, although it seemed so hitherto, but only an intuition of the mind, which may be imperfect and confused, as it was formerly, or else clear and distinct, as it is at present according as my attention directs itself more or less to the elements which it contains and of which it is composed.

However, I am greatly astonished when I consider the weakness of my mind, and its proneness to error. For although, without speaking, I consider all this in my own mind, yet words stop me, and I am almost led into error by the terms of ordinary language. For we say we see the same wax if it is put before us, and not that we judge it to be the same, because it has the same colour and shape: whence I would almost conclude that one knows the wax by the eyesight, and not by the intuition of the mind alone. If I chance to look out of a window on to men passing in the street, I do not fail to say, on seeing them, that I see men, just as I say that I see the wax; and yet, what do I see from this window, other than hats and cloaks, which can cover ghosts or dummies who move only by means of springs? But I judge them to be really men, and thus I understand, by the sole power of judgement which resides in my mind, what I believed I saw with my eyes.

A man who wishes to lift his knowledge above the common, must feel ashamed to seek occasions for doubting from the forms and terms of common speech. I prefer to avoid this and to go on to consider whether I conceived more evidently and perfectly what the wax is when I first saw it, and believed I knew it by means of my external senses, or at the very least by the common sense, as it is called, that is to say by the imaginative faculty, than I conceive it at present, after having more carefully examined what it is and by what means it can be known. Indeed, it would be ridiculous to have any doubt on this point. For what was there in that first perception that was distinct and evident, and which could not be perceived in the same way by the senses of the least of animals? But when I distinguish the wax from its external forms, and, just as if I had removed its garments, I consider it quite naked, it is certain that, although some error in my judgement may still be encountered, I cannot conceive of it in this way without possessing a human mind.

But finally, what shall I say of this mind, that is to say of myself? For so far I admit in myself nothing other than a mind. What shall I say of myself, I ask, I who seem to conceive so clearly and distinctly this piece of wax? Do I not know myself, not only with much more truth and certainty, but also more distinctly and clearly? For if I judge that the wax is, or exists, because I see it, certainly it follows much more evidently from the same fact that I myself am, or exist. For it may well be that what I see is not in effect wax; it may also be that I do not even have eyes with which to see anything; but it cannot be that, when I see or (which I no longer distinguish) think I see, I, who think, am nothing. Similarly, if I judge that the wax exists because I touch it, the same conclusion follows, namely, that I am. And if I judge thus because my imagination persuades me that it is so, or on account of any other cause whatever, I shall still draw the same conclusion. And what I have said here about the wax can apply to all the other things external to me.

Now, if my notion and knowledge of the wax seems to be more precise and distinct after it has become known to me not only by sight or touch, but also in many other ways, with how much greater distinctness, clarity and precision must I know myself, since all the means which help me to know and perceive the nature of wax, or of any other body, prove much more easily and evidently the nature of my mind? And so many other things besides are to be found in the mind itself, which can contribute to the clarification of its nature, that those which depend on the body, such as these mentioned here, scarcely deserve to be taken into account.

But now I have come back imperceptibly to the point I sought; for, since it is now known to me that, properly speaking, we perceive bodies only by the understanding which is in us, and not by the imagination, or the senses, and that we do not perceive them through seeing them or touching them, but only because we conceive them in thought, I know clearly that there is nothing more easy for me to know than my own mind. But, because it is almost impossible to rid oneself so quickly of a long-held opinion, I should do well to pause at this point, so that, by long meditation, I may imprint this new knowledge more deeply in my memory.

THIRD MEDITATION
Of God; That He Exists

I SHALL now close my eyes, stop up my ears, turn away all my senses, even efface from my thought all images of corporeal things, or at least, because this can hardly be done, I shall consider them as being vain and false; and thus communing only with myself, and examining my inner self, I shall try to make myself, little by little, better known and more familiar to myself. I am a thing which thinks, that is to say, which doubts, affirms, denies, knows a few things, is ignorant of many, which loves, hates, wills, does not will, which also imagines, and which perceives. For, as I noted above, although the things I perceive and imagine are perhaps nothing at all outside me and in themselves, I am nevertheless assured that these modes of thought, which I call perceptions and imaginations, in so far only as they are modes of thought, certainly reside and are found in me. And in the little that I have just said, I believe I have stated all I really know, or at least, all that up to now I have observed that I know.

Now I shall examine more closely if perhaps there is not to be found in me other knowledge that I have not observed before. I am certain that I am a thinking being; but do I not therefore likewise know what is required to make me certain of something? In this first knowledge, there is nothing except a clear and distinct perception of what I affirm which indeed would not be sufficient to assure me that my assertion is true, if it could ever happen that a thing I perceived to be thus clearly and distinctly true were found to be false. And consequently it seems to me that I can already establish as a general rule that all the things we conceive very clearly and distinctly are true.

Nevertheless, I have accepted and admitted before as very certain and manifest many things I have afterwards recognized to be doubtful and uncertain. What, then, were these things? They were the earth, the sky, the stars and all the other objects which I perceived by means of my senses. Now, what did I perceive clearly and distinctly in them? Certainly nothing more than that the ideas or the thoughts of these objects were presented to my mind. And even now I do not deny that these ideas are to be found in me. But there was something further which I affirmed and which, on account of the habit I had of believing it, I thought I perceived very clearly, although in reality I did not perceive it at all, namely, that there were things outside myself from which these ideas came, and to which they had a perfect resemblance. And it was in this that I was mistaken, or, if perhaps I were to judge aright, the cause of the truth of my judgement was not any knowledge I might have had.

But when I considered any very simple and easy matter in arithmetic and geometry, as, for example, that two and three make five, and other similar matters, did I not conceive them at least clearly enough to affirm that they were true? Indeed, if I have judged since that these things can be doubted, it has been for no other reason than because it occurred to me that a God might perhaps have given me a nature such that I might make mistakes even concerning the things which seem the most obvious to me. But every time this preconceived opinion of the sovereign power of a God presents itself to my mind, I am constrained to admit that it is easy for him, if he wishes, to arrange that I fall into error, even in things which I believe I know with the very greatest certainty. And on the other hand, every time I turn towards things which I think I conceive very clearly, I am so persuaded of their truth that I spontaneously declare: let him deceive me who may, but he shall never be able to cause me to be nothing, so long as I think that I am something, or to cause it one day to be true that I have never been, it now being true that I am, or that two and three make more or less than five, or such like things which I see clearly cannot be other than as I conceive them.

And indeed, as I have no reason to believe that there is a deceitful God, and as, moreover, I have not yet considered the reasons which prove that there is a God at all, the reason for doubting which depends only on this supposition is very slight, and, so to speak, metaphysical. But in order to be able to remove this doubt completely, I must inquire whether there is a God, as soon as the opportunity presents itself; and if I find that there is one, I must also inquire whether he can be deceitful; for without the knowledge of these two truths, I do not see that I can ever be certain of anything. And so that I may have the opportunity to examine this without interruption of the order of meditation which I have proposed for myself, (which is, to pass by degrees from the notions I find first in my mind to those I may discover afterwards), I must here divide all my thoughts into certain categories, and consider in which of these categories there is properly truth or error.

Among my thoughts, some are, as it were, the images of things, and it is to those alone that the name *idea* properly belongs; as when I represent to myself a man, a chimera, the sky, an angel or God himself. Others again have other forms; as when I will, fear, affirm or deny, I indeed conceive something as the object of the action of my mind, but I also add something else by this action to the idea that I have of the object; and of this class of thoughts, some are called volitions or affections, and the others judgements.

Now, concerning ideas, if they are considered only in themselves, and are not referred to any other thing, they cannot, strictly speaking, be false; for whether I imagine a goat or a chimera, it is no less true that I imagine the one than the other.

Nor need one fear that falsity may be found in the affections or the will; for although I may desire things that are bad, or even things which never existed, it is nonetheless true that I desire them.

Thus there only remain judgements in which I must take very great care not to be mistaken. Now, the principal and most usual mistake that occurs in them consists in my judging that the ideas which are within me are similar in conformity with the things outside me; for certainly, if I were to consider the ideas alone as certain modes or fashions of my thought, without wishing to relate them to anything outside, they could scarcely give rise to any error on my part.

Now, of these ideas, some seem to be innate, others adventitious and to come from outside and yet others to have been made and invented by me. For the faculty which I have of conceiving what is called in general a thing, or a truth, or a thought, seems to me to derive from nowhere else than my own nature; but if I now hear a noise, if I see the sun, or if I feel heat, up to now I have judged that these sensations came from certain things existing outside me; and finally it seems to me that sirens, hippogryphs and all other similar chimera are fictions and inventions of my mind. But I may even perhaps persuade myself that all these ideas are of the type which I call adventitious and which come from outside, or that they are all born with me, or that I have made them all, for I have not yet clearly discovered their true origin. And what I have to do here principally is to consider, concerning those which seem to come from certain objects outside me, what reasons there are for thinking that they are similar to these objects.

The first of these reasons is that it seems to me that this is taught me by nature; and the second, that my experience tells me that these ideas do not depend on my will; for often they come to me despite myself, as now, whether I will it or not, I feel heat, and because of this I am persuaded that this sensation or idea of heat is produced in me by something different from myself, viz. by the heat of the fire by which I sit. And I see nothing more reasonable than to judge that this extraneous object projects and imprints its likeness on to me rather than any other thing.

I must now see if these reasons are strong and convincing enough. When I say that it seems that this is taught me by nature, I understand by this word nature only a certain inclination which leads me to believe this, and not a natural light which would assure me of its truth. Now these two things are very different from each other; for I could not cast doubt on what the natural light shows me to be true, as for example it has already shown me that I could conclude that I exist from the fact that I doubt. And I do not possess any other faculty or power, for distinguishing true from false, which can teach me that what this light shows me to be true is not true, and which is equally to be trusted. But, concerning inclinations, which also seem to be natural to me, I have often observed, when it has been a question of choosing between virtues and vices, that they have led me no less to evil than to good; this is why I have no grounds for following them in matters regarding truth and error.

And as for the other reason, which is that these ideas must come from outside myself, since they are not dependent on my will, I do not find this convincing either. For just as these inclinations, of which I was speaking above, are to be found in me notwithstanding that they do not always accord with my will, so perhaps there is in me some faculty or power, even though I do not yet recognize it, able to produce these ideas without the help of any external things, and, indeed, it has always seemed to me until now that, when I am asleep, they are formed in me in this way without the help of the objects they represent. And finally, even though I agreed that they are caused by these objects, it is not a necessary corollary that they must be like them. On the contrary, I have often noticed in many cases, that there was a great difference between the object and its idea. Thus, for example, I find in my mind two quite different ideas of the sun: the one by which it appears to me extremely small, derives from the senses, and must be placed in the category of those which I said above come from outside; the other, by which it appears to me to be several times larger than the whole earth, is based upon the reasons of astronomy, is drawn, that is to say, from certain notions born with me, or formed by me in some way or other. Certainly these two ideas which I conceive of the sun cannot both resemble the same sun, and reason makes me believe that the idea which derives immediately from its appearance is the one which is most dissimilar.

All this shows me clearly enough that until now it has not been by a certain and premeditated judgement, but by a blind and rash impulse, that I have believed that there are things outside myself, and different from my being, which, by my sense organs or by whatever other means it may be, transmitted their ideas or images to me and imprinted their likenesses in me.

But there is still another way of inquiring whether, among the things the ideas of which are within me, there are any which exist outside me. That is to say that if ideas are taken in so far only as they are certain modes of thought, I see no difference or inequality between them, and all seem to come from me in the same way; but, considering them as images, of which some represent one thing and some another, it is evident that they are very different from one another. For, in truth, those which represent substances are undoubtedly something more and contain in themselves, so to speak, more objective reality, that is to say participate through representation in a higher degree of being or of perfection, than those which represent to me only modes or accidents. Moreover, the idea by which I conceive a God who is sovereign, eternal, infinite, unchangeable, all-knowing, all-powerful and universal Creator of all things outside himself, that idea, I say, has certainly more objective reality in it than those by which finite substances are represented to me.

Now it is manifest by the natural light that there must be at least as much reality in the efficient and total cause as in its effect: for whence can the effect draw its reality if not from its cause? And how could this cause communicate its reality to its effect, if it did not have it in itself?

And hence it follows, not only that nothingness cannot produce anything, but also that the more perfect, that is to say that which contains in itself more reality, cannot be a consequence and dependence of the less perfect. And this truth is not only clear and evident in the case of those effects which have the reality called actual or formal by the philosophers, but also in the case of ideas in which only the reality they call objective is considered. For example, the stone that has never yet existed, not only cannot now begin to exist, unless it be produced by something which has in itself formally, or eminently, everything that enters into the composition of a stone, in other words which contains in itself the same properties as those in the stone, or others superior to them; and heat cannot be produced in a subject which has earlier been devoid of it, unless it be by a cause which is of an order, degree or kind, at least as perfect as heat, and so on with other things. But further, the idea of heat, or of the stone, cannot be in me unless put there by some cause which contains in itself at least as much reality as I conceive to exist in the heat or in the stone. For although this cause does not transmit to my idea anything of its actual or formal reality, one must not on that account imagine that this cause be less real; but one must consider that, every idea being a work of the mind, its nature is such that it asks of itself no other formal reality than that which it receives and borrows from thought or the mind, of which it is only a mode, that is to say, a manner or way of thinking. Now, in order that an idea may contain one particular objective reality rather than another, it must undoubtedly receive it from some cause, in which is to be found at least as much formal reality as this idea contains objective reality. For if we suppose that something is to be found in the idea which is not to be found in its cause, this must then come from nothing; but, however imperfect this mode of existence may be by which a thing is objectively or by representation in the understanding through its idea, certainly one cannot nevertheless say that this mode of existence is nothing nor, consequently, that this idea is derived from nothing. I must not doubt either that it is necessary that reality be formally in the causes of my ideas, although the reality which I consider in these ideas be only objective, nor think it sufficient that this reality be found objectively in their causes; for, just as the mode of existing objectively belongs to ideas, by their own nature, so also the mode of existing formally belongs to the causes of these ideas (at the very least to the first and principal) by their own nature. And although one idea may give birth to another, nevertheless there cannot be an infinite regression, but one must eventually reach a first idea, whose cause is, as it were, a pattern or original, in which all the reality, or perfection, which is found only objectively, or by representation in these ideas, is contained formally and in act. Thus the natural light teaches me clearly that ideas are in me as pictures or images, which may in truth easily fall short of the perfection of the things from which they have been drawn, but which cannot ever contain anything greater or more perfect.

And the longer and more carefully I examine all these things, the more clearly and distinctly I know that they are true. But what then should I conclude from all this? It is this: if the objective reality of any one of my ideas is such that I know clearly that it is not within me, either formally or eminently, and that consequently I cannot myself be its cause, it follows necessarily from this that I am not alone in the world, but that there is besides some other being who exists, and who is the cause of this idea; whereas, if no such idea is to be found within me, I would have no argument which could convince me and make me certain of the existence of any other being except myself, for I have carefully examined all these arguments and I have not been able to find any other up to now.

Now among these ideas, besides that which represents me to myself, about which there can be no difficulty here, there is one which represents to me a God; others corporeal and inanimate things, others angels, others animals, and finally some which represent to me men like myself. But concerning the ideas which represent to me other men, or animals, or angels, I easily conceive that they may be formed by the mixture and composition of the other ideas I have of corporeal things and of God, even if, outside myself, there were no other men in the world, no animals, and no angels. And concerning the ideas of corporeal things, I recognize nothing so great, nor so excellent there which seems incapable of coming from me; for, if I look at these ideas more closely, and if I examine them in the same way as I yesterday examined the idea of wax, I find very few things there that I perceive clearly and distinctly: viz. magnitude or extension in length, width and depth; figure, which is formed by the limits and boundaries of extension; position, which bodies of different figure maintain among themselves, and movement or the change of position; to which can be added substance, duration and number. As for the other things, such as light, colour, sound, smell, taste, heat, cold, and the other tactile qualities, they are to be found in my thought with such obscurity and confusion that I do not know even whether they are true, or false and only apparent, that is to say, whether the ideas I form of these qualities are truly the ideas of real things, or whether they represent to me only chimera which cannot exist. For, although I observed above that it is only in judgements that proper formal falsity can be met with, nevertheless a certain material falsity can be found in ideas, when they represent what is nothing as if it were something. For example, the ideas I have of cold and heat are so unclear and indistinct that I cannot discern from them if cold is only a privation of heat, or if heat is a privation of cold; or if they are both real qualities or not; and since, ideas being as it were images, there can be none which do not seem to us to represent something, if it is true to say that cold is nothing other than a privation of heat, the idea which represents it to me as something real and positive, will not unaptly be called false, and the same may be said of other similar ideas, to which indeed it is not necessary for me to attribute any other author than myself. For, if they are false, that is to say, if they represent things which are not, the natural light tells me that they originate from nothing, that they are within me only because something is lacking in my nature, because my nature is not all perfect. But if these ideas are true, yet because they exhibit to me so little reality that I cannot even distinguish clearly the object represented from non-being, I see no reason why they should not be produced by me, and why I should not be the author of them.

As for the clear and distinct ideas I have of corporeal things, there are some which it seems I may have drawn from the idea I have of myself, such as the idea of substance, duration, number and other similar things. For when I think that a stone is a substance, or a thing which is capable of existing of itself, that I also am a substance, although I see clearly that I am a thinking and non-extended thing, and that the stone, on the contrary, is an extended non-thinking thing, and that there is thus a notable difference between these two concepts, all the same they seem to come together in that they both represent substances. Similarly, when I think that I exist now, and recollect besides that I existed in the past, and when I conceive various thoughts, the number of which I know, then I acquire the ideas of duration and number which I can thereafter transfer to all the other objects I wish.

As for the other qualities of which the ideas of corporeal things are composed, namely, extension, figure, position and movement, it is true that they are not formally in me, since I am only a thinking thing; but because these are only certain modes of substance, and, as it were, the clothing under which corporeal substance appears to us, and because I myself also am a substance, it seems that they may be contained in me eminently.

There remains, then, only the idea of God, in which I must consider whether there is anything which could not have come from me. By the name of God I understand an infinite substance, eternal, immutable, independent, omniscient, omnipotent, and by which I and all the other things which exist (if it be true that any such exist) have been created and produced. But these attributes are so great and eminent, that the more attentively I consider them, the less I am persuaded that the idea I have of them can originate in me alone. And consequently I must necessarily conclude from all I have said hitherto, that God exists; for, although the idea of substance is in me, for the very reason that I am a substance, I would not, nevertheless, have the idea of an infinite substance, since I am a finite being, unless the idea had been put into me by some substance which was truly infinite.

And I must not imagine that I do not conceive the infinite by means of a true idea, but only by the negation of the finite, in the same way as I comprehend rest and darkness by the negation of movement and light: for, on the contrary, I see manifestly that there is more reality in the infinite substance than in the finite, and hence that I have in me in some way the notion of the infinite, before that of the finite, that is to say the notion of God, before that of myself. For how would it be possible for me to know that I doubt and desire, that is to say, that I lack something and am not all perfect, if I did not have in me any idea of a more perfect being than myself, by comparison with which I know the deficiencies of my nature?

And one cannot say that this idea of God is perhaps materially false, and consequently that I can obtain it from nothing, that is to say, that it can be in me because I am imperfect, as I have said above of the ideas of heat and cold and other similar things: for, on the contrary, this idea, being very clear and distinct, and containing in itself more objective reality than any other, there is no other which can be of itself so true, or which may be less suspected of error and falsity.

The idea, I say, of this supremely perfect and infinite being is entirely true; for although perhaps one may pre-

tend that such a being does not exist, one cannot nevertheless pretend that the idea of him represents nothing
real, as I have already said of the idea of cold.

This same idea is also very clear and very distinct, since
whatever my mind conceives clearly and distinctly as real
and true and containing in itself any perfection, is contained and enclosed entirely in this idea.

And this does not cease to be true, although I do not
understand the infinite, or although there be in God an
infinity of things that I cannot understand, nor perhaps
even reach in any way by thought; for it is in the nature
of the infinite that it should not be understood by my
nature, which is finite and restricted; and it is sufficient
that I understand this correctly, and that I judge that all
the things I perceive clearly and in which I know there is
some perfection, and perhaps also an infinity of properties of which I am ignorant, are in God formally or eminently, in order that the idea I have of him may be the
most true, clear and distinct of all the ideas which are in
my mind.

But perhaps also I am something more than I imagine
myself to be, and all the perfections I attribute to the
nature of a God are in some way potentially in me, although they have not yet been realized and do not manifest themselves in actions. In truth, experience tells me
already that my knowledge increases and becomes more
perfect little by little, and I see nothing to prevent it
increasing more and more up to infinity; then, my knowledge being thus increased and perfected, I see nothing to
prevent me from acquiring thereby all the other perfections of the divine nature; and finally it seems that the
power I have to acquire these perfections, if it is really in
me, should be sufficient to produce the ideas of them. Yet,
on looking into the matter more closely, I discover that
this cannot be; for, firstly, although it were true that my

knowledge acquired daily new degrees of perfection, and although there were potentially in my nature many things which are not yet there actually, still, all these excellences do not belong to or approach in any way the idea I have of the Divinity, in whom nothing is to be found only potentially but all actually existent. And is it not even an infallible argument of the existence of imperfection in my knowledge that it grows little by little and increases by degrees? Moreover, although my knowledge increases more and more, nevertheless I know well enough that it will never be actually infinite, for it will never reach so high a point of perfection that it will not still be capable of further increase. But I conceive God as actually infinite in so high a degree that nothing can be added to the sovereign perfection which he possesses. And finally I see full well that the objective being of an idea cannot be produced by a being which exists only potentially which, properly speaking, is nothing, but only by a being existing formally or actually.

And, in truth, I see nothing in all that I have just said which is not very easy for all, who give it careful thought, to know by the natural light of the mind; but when I let my attention relax somewhat, my mind, finding itself obscured and, as it were, blinded by the images of sensible objects, does not readily remember the reason why the idea I have of being more perfect than I am, must necessarily have been put in me by a being which is in reality more perfect.

This is why I wish here to go further, and to inquire whether I, who have this idea of God, could exist, if there were no God. And I ask, from whom could I then derive my existence? Perhaps from myself, or from my parents, or from some other causes less perfect than God, for one cannot imagine anything more perfect or even equal to him.

Now, if I were independent of all other existence, and were myself the author of my being, I should certainly doubt nothing, I should conceive no desires, and finally I would lack no perfection; for I should have given myself all those perfections of which I have in me some idea, and thus I should be God.

And I must not imagine that the things I lack are perhaps more difficult to acquire than those which I already possess for, on the contrary, it is quite clear that it was much more difficult that I, that is to say a thing or a substance which thinks, should proceed from nothing, than it would be for me to acquire the knowledge of many things of which I am ignorant and which are only the accidents of a thinking substance. And thus, if I had given myself this greater perfection of which I have just spoken, that is to say, if I were the author of my birth and existence, I would not at least have denied myself things which are more easily acquired, viz. much of the knowledge which I now lack. Neither would I have denied myself any of the attributes which are contained in the idea I conceive of God, because there is none of these which seems to me more difficult to acquire; and if there were any such, they would certainly appear so to me (assuming that all the others I possess came from me), for I should discover that my power ended with them and was incapable of reaching them.

And although I were to suppose that I have always been as I am now, I could not, on that account, escape the force of this reasoning, and fail to be aware that God is necessarily the author of my existence. For the whole time of my life may be divided into an infinity of parts, each of which depends in no way on the others; and thus, it does not follow that because I existed a little earlier, I must exist now, unless at this moment some cause produces and creates me anew, so to speak, that is to say, conserves me.

In truth, it is quite clear and evident to all those who will attentively consider the nature of time, that a substance, in order to be conserved in each moment of its duration, needs the same power and action that would be necessary to produce and create it afresh, if it did not yet exist. So that the natural light shows us clearly that conservation and creation differ only in regard to our mode of thinking, and not at all in fact. It is only necessary therefore for me to ask myself if I possess any power or virtue capable of acting in such a way that I, who exist now, shall still exist in the future: for, since I am nothing but a thinking thing (or, at least, since up to now only precisely this part of me is concerned), if such a power resided in me, indeed I should at the very least be conscious of it; but I am conscious of no such power and, thereby, I know evidently that I depend on some being different from myself.

But perhaps that being on whom I depend is not what I call God, and I am produced either by my parents, or by some other causes less perfect than God. Far from it, for, as I have already said, it is very obvious that there must be at least as much reality in the cause as in its effect. And accordingly, since I am a thinking thing, and have in me an idea of God, whatever finally the cause may be to which my nature is attributed, it must necessarily be admitted that the cause must equally be a thinking thing, and possess within it the idea of all the perfections that I attribute to the divine nature. Then one may inquire whether this cause owes its origin and its existence to itself, or to some other thing. For if it owes it to itself, it follows, from the reasons I have advanced above, that it must be God; for, having the virtue of being and existing of itself, it must also without doubt have the power of actually possessing all the perfections of which it conceives the idea, that is to say, all those I conceive to be in

God. But if it owes its existence to some cause other than itself, we shall ask again, for the same reason, whether this second exists of itself, or through some other, until, by degrees, we arrive finally at an ultimate cause which will be God. And it is quite manifest that in this matter there can be no infinite regress seeing that it is not a question here so much of the cause which once produced me as of that which conserves me at this moment.

Nor can we pretend that perhaps several causes together had a hand in my production, and that from one I received the idea of one of the perfections I attribute to God, and from another the idea of some other, with the result that all these perfections are indeed to be found somewhere in the universe, but do not meet all together to be assembled in one single being who is God. For, on the contrary, the unity, simplicity or inseparability of all the properties of God, is one of the principal perfections that I conceive to be in him; and, indeed, the idea of this unity and assembly of all the perfections of God cannot have been put in me by any cause from which I did not also receive the ideas of all the other perfections. For it cannot have made me conceive of them as united and inseparable without at the same time having given me to know, in some sense, what they were and acquainted me of their existence in a particular mode.

As for my parents, from whom it appears that I derive my birth, although all that I have ever believed about them be true, nevertheless this does not mean that it is they who conserve me, or who made me and produced me in so far as I am a thinking thing, since all they did was to put certain dispositions into this matter in which I judge that I, that is to say my mind, which alone I take now as being myself, is enclosed; and thus there can be no difficulty here in respect of them, and we must necessarily conclude that, from the mere fact that I exist, and that the idea of a sovereignly perfect being, that is to say of God, is in me, the existence of God is very clearly demonstrated.

It remains for me only to examine how I have acquired this idea; for I have not received it through the senses, nor has it ever presented itself to me unexpectedly, as do the ideas of sensible objects, when these things present themselves or seem to present themselves to the external organs of my senses. Neither is it a pure production or fiction of my mind, for it is not in my power to take away from or to add to it. And consequently there remains nothing more to say, except that, as with the idea of myself, it was born and produced with me at the moment of my creation.

And, in truth, it is not to be thought strange that God, in creating me, should have put in me this idea to serve, as it were, as the mark that the workman imprints on his work; nor is it necessary that this mark should be something different from the work itself. But, from the mere fact that God created me, it is highly credible that he in some way produced me in his own image and likeness, and that I perceive this likeness, in which the idea of God is contained, by means of the same faculty by which I apprehend myself; that is to say that, when I reflect upon myself, not only do I know that I am an imperfect, incomplete and dependent being, and one who tends and aspires unceasingly towards something better and greater than I am, but know also, at the same time, that he upon whom I depend possesses in himself all the great attributes to which I aspire, and the ideas of which I find in me, not merely indefinitely and potentially, but actually and infinitely, and that he is thus God. And the whole force of the argument I have used here to prove the existence of God consists in this, that I recognize that it would not be possible for my nature to be as it is, that is to say, that I should have in me the idea of a God, if God did not really exist; this same God, I say, the idea of whom is in me, that is to say, who possesses all these high perfections of which the mind may well have some idea without, however, being able fully to understand them, who is not subject to any defects, and who has none of the things which indicate some imperfection. Whence it is clear enough that he cannot be a deceiver, for the natural light teaches us that deceit stems necessarily from some defect.

But, before I examine this more closely, and pass on to the consideration of other truths which can be derived from it, it seems to me very appropriate to pause for a time in contemplation of this all-perfect God, to ponder at leisure his marvellous attributes, to consider, admire and adore the incomparable beauty of this immense light, as far, at least, as the strength of my mind, which is, so to speak, dazzled by it, will permit.

For, just as faith teaches us that the sovereign felicity of the other life consists in the contemplation of the divine Majesty alone, so even now we can learn from experience that a similar meditation, although incomparably less perfect, allows us to enjoy the greatest happiness we are capable of feeling in this life.

FOURTH MEDITATION

Of Truth and Error

I HAVE so accustomed myself these past days to detach my mind from the senses, and I have so accurately observed that there are very few things one can know with certainty about corporeal objects, that there are many more things which are known to us about the human mind, and many more still about God himself, that I shall now turn my mind away without difficulty from the consideration of sensible or imaginable things, in order to bring it to bear on those which, being disengaged from all matter, are purely intelligible.

And certainly the idea I have of the human mind, inasmuch as it is a thinking thing, and not extended in length, breadth and depth, and does not participate in anything that pertains to the body, is incomparably more distinct than the idea of any corporeal object. And when I consider that I doubt, that is to say, that I am an incomplete and dependent being, the idea of a complete and independent being, that is to say of God, presents itself to my mind with such distinctness and clearness, and, from the fact alone that this idea is found in me, or that I, who possess this idea, am or exist, I conclude so evidently that God exists, and that my existence depends entirely on him in each moment of my life, that I do not think that the human mind can know anything with more clearness and certainty. Already, then, I seem to discover a path that will lead us from the contemplation of the true God, in whom all the treasures of knowledge and wisdom are contained, to the knowledge of the other things in the universe.

For, in the first place, I recognize that it is impossible that he should ever deceive me, since in all fraud and deceit is to be found a certain imperfection; and although it may seem that to be able to deceive is a mark of subtlety or power, yet the desire to deceive bears evidence without doubt of weakness or malice, and, accordingly, cannot be found in God.

Secondly, I am aware in myself of a certain power of judgement, which undoubtedly I have received from God, in the same way as all the other things which I possess; and as he would not wish to deceive me, it is certain that he has not given to me a power such that I can ever be in error, if I use it properly. And there would remain no doubt of this truth, if one could not, it seems, draw from it the conclusion that, in consequence, I can never be mistaken; for, if I owe everything I possess to God, and if he has not given me power in order to fall into error, it seems that I can never be mistaken. And, in truth, when I think only of God, I discover in myself no cause of error or falsehood; but then afterwards, returning to myself, experience tells me that I am nevertheless subject to an infinity of errors. And seeking the cause of these more closely, I observe that it is not only a real and positive idea of God or of a supremely perfect being which presents itself to my mind, but also, so to speak, a certain negative idea of nothing, that is to say of that which is infinitely distant from all sort of perfection; and that I am, as it were, midway between God and nothing, or placed in such a way between the supreme being and non-being, that there is, in truth, nothing in me which can lead me into error, in so far as a sovereign being has produced me; but that if I consider myself as participating in some way in nothing or non-being, that is to say in so far as I am not myself the sovereign being, I find myself exposed to an infinity of deficiencies, so that I must not be surprised if I make mistakes.

Thus I discern that error, as such, is not something real

which depends on God, but that it is simply a defect; and accordingly, that in order to fall into error I do not need some power given me specially by God to this end, but that my being mistaken arises from the fact that the power which God has given me of discerning the true from the false is not infinite in me.

However, this does not yet altogether satisfy me; for error is not a pure negation, that is to say, is not the simple deficiency or lack of some perfection which is not due to me, but rather it is a deprivation of some knowledge which it seems I ought to possess. And, considering the nature of God, it does not seem possible to me that he should have given me any faculty not perfect in its kind, that is to say, which lacks some perfection due to it; for if it is true that the more expert the worker, the more perfect and accomplished the works which come from his hands, what being shall we imagine has been produced by the supreme Creator of all things, that is not perfect and complete in all its parts? And assuredly there is no doubt that God could have created me in such a way that I should never fall into error; it is certain too that he always wills what is best: is it more advantageous then, for me to be deceived than not to be deceived?

Considering this more attentively, the first thought that occurs to me is that I must not be surprised if my intelligence is not capable of understanding why God does what he does, and that thus I have no reason to doubt his existence merely because I see perhaps from experience many other things without being able to understand why or how God has produced them. For, knowing already that my nature is extremely weak and limited, and that God's nature, on the contrary, is immense, incomprehensible and infinite, I no longer have difficulty in recognizing that there is an infinity of things in his power, the causes of which are beyond the range of my mind.

And this reason alone is sufficient to persuade me that the whole class of final causes is of no use in physical or natural things; for it does not seem to me that I can, without temerity, seek to discover the impenetrable ends of God.

In addition it further occurs to me, that one must not consider a single creature separately, when one seeks to inquire into the perfection of God's works, but generally all creatures together. For the same thing that might, perhaps, with some reason, seem very imperfect if quite alone, may be very perfect in its nature if it is looked upon as part of the whole universe. And although, since I resolved to doubt everything, I have known with certainty only my own existence, and that of God, nevertheless, after having recognized the infinite power of God, I cannot deny that he may have produced many other beings, or at least that he can produce them, so that I exist and occupy a place in the world as a part of the total number of beings in the universe.

Whereupon, examining myself more closely, and considering what my errors are which alone bear witness to the existence of imperfection in me, I see that they depend on the concurrence of two causes, namely, the power I have of knowing things, and the power of choice, or free will, that is to say, of my understanding, and of my will. For by understanding alone I neither affirm nor deny anything, but merely conceive the ideas of things, which I can affirm or deny. Now, in considering it thus precisely, it can be said that there is never to be found any error in it provided that one takes the word error in its proper signification. And although there may perhaps be an infinity of things in the world of which I have no idea in my understanding, it cannot be said on that account that my understanding is deprived of these ideas, as of something which is due to its nature, but only that it does not have them, because, in truth, there is no reason which can prove that God should have given me a greater and more ample faculty of knowing than he actually has; and, however skilful and accomplished a workman I imagine him to be, I must not, on that account, think that he should have put into each of his works all the perfections he can put in some. Nor, moreover, can I complain that God has not given me a free will, or a will sufficiently ample and perfect, since, indeed, I am conscious of possessing a will so ample and extended as not to be enclosed in any limits. And what seems to me here to be very remarkable is that, of all the other faculties in me, there is none so perfect and extended that I do not clearly conceive that it could be even greater and more perfect. For, to give an example, if I consider the faculty of understanding which is in me, I find that it is of a very small extent, and greatly limited, and at the same time form the idea of another similar faculty, much more ample and even infinite; and from the mere fact that I can form the idea of it, I know without difficulty that it pertains to the nature of God. In the same way, if I examine my memory, or imagination, or any other faculty, I find none which is not very small and limited, and in God immense and infinite. It is will alone that I experience to be so great in me that I conceive the idea of no other as more ample and more extended; so that it is my will principally which tells me that I bear the image and resemblance of God. For, although the will is incomparably greater in God than in me, both from the point of view of the knowledge and power, which, being joined with it, make it stronger and more efficacious, and of its object inasmuch as it extends to infinitely more things, nevertheless it does not appear to me greater, if I consider it in itself formally and precisely. For the power of the will consists only in our being able to do a thing, or not to do it (that is to say, to affirm or deny, to pursue or to flee) or rather only, when affirming or denying, pursuing or fleeing the things our understanding proposes to us, in our acting in such a way that we are not conscious that any external force is constraining us. For, in order that I may be free, it is not necessary for me to be indifferent to the choice between one or other of two contraries; but rather, the more I lean towards one, either because I know clearly that the good and the true are in it, or because God so disposes my mind from within the more freely do I make my choice and embrace it. And indeed divine grace and natural knowledge, far from diminishing my freedom, increases it rather, and strengthens it. So that the indifference I feel when I am not inclined to one side rather than to another by the weight of any reason, is the lowest degree of freedom, and reveals a defect of knowledge rather than a perfection of will; for if I always knew clearly what is true and good, I should never have difficulty in deciding which judgement and which choice I should make, and thus I should be entirely free without ever being indifferent.

From all this I perceive that neither the power of willing which I have received from God is in itself the cause of my errors, for it is very ample and perfect of its kind; nor is the power of understanding or conceiving, for, conceiving nothing except by means of the faculty which God has given me, without doubt all that I conceive I conceive correctly, and it is impossible for me to be deceived in it. Whence, then, arise my errors? From this fact alone, that the will being much more ample and extended than the understanding, I do not contain it within the same limits, but extend it also to things I do not understand, and the will being of itself indifferent to such things, very easily goes astray and chooses the bad instead of good, or the false instead of the true, which results in my falling into error or sinning.

For example, when inquiring these last few days whether anything existed in the world, and finding that, from the very fact that I was examining this question, it followed most clearly that I existed myself, I could not help judging that what I conceived so clearly was true; not that I was forced to this conclusion by any external cause, but simply because the great clarity of my understanding was followed by a great inclination in my will; and I was led to believe with all the more freedom as I was the less indifferent. On the other hand, at the moment I not only know I exist, inasmuch as I am a thinking being, but also a certain idea of corporeal nature is presented to my mind which makes me doubt whether the thinking nature which is in me, or rather by which I am what I am, is different from this corporeal nature, or whether both are merely one and the same thing. And here I suppose that I do not yet know of any reason which persuades me one way or the other; from which it follows that it is a matter of complete indifference to me which of the two conclusions to deny or to affirm or even to abstain from giving any judgement at all on the matter.

And this indifference extends not only to things of which the understanding has no knowledge, but generally also to all those which it does not discover with perfect clarity at the moment the will is deliberating on them; for, however probable the conjectures may be which make me inclined to form a judgement on something, the mere knowledge that these are only conjectures and not certain and indubitable reasons, suffices to cause me to judge the opposite. This is something which I have experienced sufficiently during the last few days, when I laid down as false all I had held hitherto to be very true, merely because I noticed that one could have some doubts about it.

But if I abstain from giving my judgement on a thing when I do not conceive it clearly and distinctly enough, it is evident that I act rightly and am not deceived; but if I decide to deny or affirm it, then I no longer make use as I should of my free will; and if I affirm what is false, it is evident that I am deceived, and even though I judge according to the truth, it is only by chance, and I am none the less at fault and misuse my free will; for the natural light teaches us that the knowledge of the understanding must always precede the determination of the will. And it is in this wrong use of free will that is found the privation which constitutes the form of error. Privation, I say, is found in the operation, in so far as it comes from me, but it is not in the faculty that I have received from God, nor in the operation, in so far as it depends on him. For I have indeed no cause to complain that God has not given me a greater power of intelligence, or a greater natural light than he has, for it is of the nature of a finite understanding not to understand an infinity of things, and of the nature of a created understanding to be finite; but I have every reason to thank him, in that, having never owed me anything, he has nevertheless given me the few perfections which I possess; and I am far indeed from conceiving such unjust sentiments as to imagine that he has unjustly taken away or withheld the other perfections he has not given me. I have no reason, moreover, to complain of his having given me a will more extensive than my understanding, since, as the will consists of only one and, as it were, an indivisible thing, it appears that it is of such a nature that nothing could be taken away from it without destroying it; and indeed the more extensive it is, the more I should thank the goodness of him who gave it to me. And finally also I must not complain that God concurs with me in forming the acts of the will, that is to say the judgements, in which I am in error, because those acts are entirely true and absolutely good, in so far as they depend on God; and there is, as it were, more perfection in my nature in that I am able to form them, than if I were not. As for privation, in which alone consists the formal reason of error and sin, it has no need of the concurrence of God, since it is not a thing or a being, and if it is referred back to God as to its cause, it should not be called privation, but only negation, according to the meaning given to these words by the Schoolmen.

For, in truth, it is not an imperfection in God, that he has given me the freedom to give my judgement, or to withhold it, concerning things of which he has not put a clear and distinct knowledge into my understanding; but undoubtedly it is an imperfection in me that I do not use it well, and give my judgement rashly on things which I perceive only obscurely and confusedly.

I see, nevertheless, that it would have been easy for God to have arranged matters in such a way that I should never be deceived, although I still remained free, and with a limited knowledge, namely, by giving to my understanding a clear and distinct knowledge of all the things on which I should ever have to deliberate, or merely by so deeply engraving in my memory the resolution never to judge anything without conceiving it clearly and distinctly, that I should never forget it. And I well observe that inasmuch as I consider myself alone, as if there were only me in the world, I should have been much more perfect than I am, if God had created me in such a way that I never fell into error. But I cannot therefore deny that it is not in some way a greater perfection in the universe taken as a whole that some of its parts are not exempt from defect, than if they are all alike. And I have no right to complain if God, having put me in the world, has not wished to put me at the level of the most noble and perfect things; indeed, I have reason to be satisfied that, if he has not given me the virtue of not falling into error by the first means that I set out above, which depends on a clear and evident knowledge of everything on which I can deliberate, he has at least left in my power the other means, which is to retain firmly the resolve never to give my judgement on things the truth of which is not clearly known to me. For although I know this weakness in my nature which makes it impossible for me to keep my mind continually fixed on the same thought, I can all the same, by attentive and oft-repeated meditation, imprint it so strongly in my memory that I never fail to recall it every time I have need to, and I can acquire in this way the habit of not falling into error. And, in so far as it is in this that the greatest and principal perfection of man consists, I consider that I have gained not a little by this Meditation, in having discovered the cause of falsity and error.

And certainly there can be no other cause than that which I have explained; for every time that I so restrain my will within the limits of my knowledge that it makes no judgement except on things which are clearly and distinctly represented to it by the understanding, I cannot be in error; because every clear and distinct conception is undoubtedly something real and positive, and therefore cannot originate from nothing, but must necessarily have God as its author, God, I say, who, being supremely perfect, cannot be the cause of any error; and consequently one must conclude that such a conception or such a judgement is true.

Besides, I have not only learnt today what I must avoid in order to escape error, but also what I must do in order to arrive at knowledge of the truth. For certainly I shall reach truth if I fix my attention sufficiently on all the things I conceive perfectly, and if I separate them from others which I apprehend only confusedly and obscurely, which, from now on, I shall take great care to do.

FIFTH MEDITATION

*Of the Essence of Material Things; and, Once More
of God, that He Exists*

THERE remain several other things for me to consider regarding the attributes of God, and my own nature, that is to say my mind, but I shall perhaps resume the investigation of these at another time. Now, having observed what must be done or avoided to arrive at knowledge of the truth, what I have principally to do is to try to emerge from and rid myself of all those doubts into which I fell during the last few days, and to see if anything can be known for certain about material things.

But before considering whether such things exist outside myself, I must examine the idea I have of them in so far as they are to be found in my thought and see which of them are distinct and which are confused.

In the first place, I imagine distinctly that quantity which the philosophers commonly call continuous or the extension in length, breadth and depth which is in this quantity or rather in the thing to which it is attributed. Further, I can enumerate in it several diverse parts, and attribute to each of these parts all sorts of sizes, figures, situations and movements; and finally, I can assign to each of these movements all sorts of duration.

And I not only know these things distinctly when I consider them in general, but also, in so far as I apply my attention, I conceive an infinity of particulars concerning numbers, figures, movements, and so on, the truth of which appears so evidently and accords so well with my nature that when I begin to discover them it does not seem that I am learning anything new, but rather recalling what I already knew before, that is to say, perceiving things which were already in my mind, although I had not yet turned my thoughts towards them.

And what I find here of greatest importance is that I find in myself an infinity of ideas of certain things which cannot be considered as pure nothing, although perhaps they have no existence outside my thought, and which are not invented by me, although it may be in my power to think or not to think them, but possess their own true and immutable natures. As, for example, when I imagine a triangle, although there may not perhaps be, and never has been, any place in the world outside my thought such a figure, yet it remains true that there is a certain determined nature or form or essence of this figure, immutable and eternal, which I have not invented, and which does not depend in any way on my mind. This is apparent from the fact that it is possible to demonstrate diverse properties of the triangle, viz. that its three angles are equal to two right-angles, that the greatest angle is suspended by the greatest side, and so on, which now, whether I will it or not, I recognize very clearly and very evidently to belong to it, although I had not hitherto thought of them in any way, when I imagined a triangle for the first time; and consequently it cannot be said that I have invented them.

Nor is there any point in the objection that perhaps this idea of a triangle has come into my mind by the intermediary of the senses, since I have sometimes seen bodies of triangular shape; for I can form in my mind an infinity of other figures of which not the least suspicion can be had that they have ever come within the scope of my senses, and I can nevertheless demonstrate diverse properties of their nature, just as well as that of the triangle, all of which must indeed be true, since I conceive them clearly. And consequently they are something and not a mere nothing; for it is very evident that all that is true is something, and I have already amply shown above that all the things I know clearly and distinctly are true. And even though I had not demonstrated it, the nature of my mind is such that I could not help considering them to be true, while I perceive them clearly and distinctly. And I remember that, even when I was still strongly attached to the objects of the senses, I reckoned among the most constant truths those I conceived clearly and distinctly relating to figures, numbers and other things which pertain to arithmetic and geometry.

And now, if from the fact alone that I can draw from my thought the idea of a thing, it follows that all that I recognize clearly and distinctly as belonging to that thing does indeed belong to it, cannot I derive from this an argument and a proof demonstrating the existence of God? It is certain that I no less find the idea of God in me, that is to say, the idea of a supremely perfect being, than that of any figure or number whatsoever. And I know no less clearly and distinctly that an actual and eternal existence belongs to his nature, than that all I can demonstrate of any figure or number truly belongs to the nature of that figure or number. And consequently, even though everything I have concluded in the preceding Meditations were shown to be untrue, the existence of God must pass in my mind for at least as certain as I have until now judged all the truths of mathematics to be, which concern only numbers and figures, although indeed this may not appear at first entirely manifest, and may seem to have the appearance of a sophism. For being accustomed in all other matters to distinguish between existence and essence, I am easily persuaded that the existence of God can be separated from his essence and that, thus, God may be conceived as not actually existing. But, nevertheless, when I think about it more attentively, it becomes manifest that existence can no more be separated from the essence of God than the fact that the sum of its three angles is equal to two right-angles can be separated from the essence of a triangle or than the idea of a mountain can be separated from the idea of a valley; so that there is no less contradiction in conceiving a God, that is to say, a supremely perfect being, who lacks existence, that is to say, who lacks some particular perfection, than in conceiving a mountain without a valley.

But although, in truth, I cannot conceive a God without existence, any more than I can a mountain without a valley, yet, just as it does not follow that merely because I conceive a mountain with a valley, there is any mountain in the world, so similarly, although I conceive God as having existence, it does not follow from that, that there is a God who actually exists, for my thought imposes no necessity on things; and as I can well imagine a winged horse, although there is no such horse, so could I perhaps attribute existence to God, even though no God existed. But this is fallacious and there is here a sophism under this seeming objection: for, because I cannot conceive a mountain without valley, it does not follow that there is any mountain in the world, or any valley, but only that the mountain and the valley, whether they exist or not, cannot in any way be separated from one another; whereas, because I cannot conceive God without existence, it follows that existence is inseparable from him, and hence that he truly exists; not that my thought can make this be so, or that it imposes any necessity on things, but, on the contrary, because the necessity of the thing itself namely, the necessity of the existence of God, determines my thought to conceive in this way. For I am not free to conceive a God without existence, that is to say, a supremely perfect being devoid of a supreme perfection, as I am free to imagine a horse with or without wings.

Nor must it be alleged here that it is in truth necessary

to admit that God exists, after having supposed that he possesses all perfection, since existence is one of them, but that my first supposition was itself unnecessary; in the same way that it is not necessary to think that all four-sided figures can be inscribed in the circle, for if I supposed this, I must perforce admit that the rhombus can be inscribed in the circle, since this is a four-sided figure; and thus I would be constrained to assert something false. But this argument is not valid: for although it may not be necessary that I ever have any thought of God, nevertheless, every time I happen to think of a first and sovereign being, and, so to speak, to draw the idea of him from the treasure-house of my mind, I must necessarily attribute to him all sorts of perfections, although I cannot manage to enumerate them all or apply my mind to each one individually. And this necessity is sufficient, after I have recognized that existence is a perfection, to make me conclude that this first and sovereign being truly exists in the same way that it is not necessary ever to imagine any triangle, but every time I wish to consider a rectilineal figure composed of only three angles, it is absolutely necessary that I should attribute to it all those properties which lead one to conclude that its three angles are not greater than two right-angles, although perhaps I may not then consider that fact in particular. But, when I consider which figures are capable of being inscribed in the circle, it is not necessary in any way for me to think that all four-sided figures are of this number; on the contrary, I cannot even pretend that it is so, so long as I am unwilling to accept anything into my thought except what I can conceive clearly and distinctly. And consequently, there is a big difference between false suppositions, like this one, and true ideas which are born with me, the first and chief of which is the idea of God.

For, indeed, I recognize in several ways that this idea is

not something factitious, depending only on my thought, but that it is the representation of a true and immutable nature: firstly, because I cannot conceive of anything other than God alone, to whose essence existence belongs of necessity; secondly, because it is impossible to conceive of two or several Gods of the same kind. And, given that there is one now who exists, I see clearly that he must necessarily have existed from all eternity, and must exist eternally in the future. And finally, because I know an infinity of other attributes in God, none of which I can diminish or change.

Moreover, whatever proof and argument I use, it must always come back to this, that only the things I conceive clearly and distinctly have the power to convince me completely. And although, among the things I conceive in this way, there are indeed some which are obvious to everyone, while others reveal themselves only to those who consider them more closely and examine them more precisely, nevertheless, after they have once been discovered, the latter are not considered less certain than the former. Thus, for example, in every right-angled triangle, although it is not at first so easily perceived that the square of the base is equal to the squares of the other two sides, as that this base is opposite to the greatest angle, nevertheless, once this is recognized, we are equally persuaded of the truth of the former as of the latter. And, as regards God, if my mind were not already obscured by prejudices, and my thought distracted by the continual presence of the images of sensible objects, there would be nothing that I should know sooner or more easily than him. For is there anything of itself clearer and more manifest than the existence of a God, that is to say a supreme and perfect being, in the idea of whom alone is contained necessary or eternal existence, and who, consequently, exists?

And although, in order to conceive this truth clearly, I have needed great application of mind, nevertheless I consider myself not only as assured of it as of everything which seems to me most certain, but in addition, I see that certainty concerning all the other truths depends on it so absolutely that, without this knowledge, it is impossible ever to know anything perfectly.

For although I am of such a nature that, as soon as I understand something very clearly and distinctly, I am naturally inclined to believe it to be true, nevertheless, because I am also of such a nature that I cannot keep my mind always fixed on the same object, and as I often remember having judged a thing to be true, when I have ceased to have in mind the reasons which obliged me so to judge, it may happen meanwhile that other reasons are presented to me which would easily make me change my opinion, if I did not know that God existed; and thus I would never have a true and certain knowledge of anything whatever, but only vague and shifting opinions.

Thus, for example, when I consider the nature of the triangle, I know clearly, having some knowledge of geometry, that its three angles are equal to two right-angles, and it is impossible for me not to believe it while I apply my mind to its proof; but as soon as I turn my mind away from the proof, even though I remember having clearly understood it, yet it may happen that I doubt the truth of it, if I do not know that there is a God. For I may persuade myself that I have been so made by nature as to be easily deceived, even in matters which I believe I understand most with the greatest evidence and certainty, especially when I remember having often considered many things to be true and certain, which afterwards other reasons have caused me to judge to be absolutely false.

But after I have discovered that God exists, recognizing

at the same time that all things depend on him, and that he is no deceiver, and consequently judged that everything I perceive clearly and distinctly cannot fail to be true, although I no longer have present in my mind the reasons for my judgement, no contrary reason can be adduced which could ever make me doubt its truth, provided that I remember that I once clearly and distinctly comprehended it. Thus I have a true and certain knowledge of it. And this same knowledge also extends to all the other things I remember having formerly proved as the truths of geometry and the like: for what objections can be brought to lead me to doubt them? Shall it be said that my nature is such that I am subject to error? But I know already that I cannot be deceived in judgements the reasons of which I know clearly. Shall it be said that I formerly held many things to be true and certain which I have later recognized to be false? But I did not know any of those things clearly and distinctly, and, not yet knowing the rule by which I am assured of the truth, I was led to believe them by reasons which I afterwards discovered to be less strong than I had at the time imagined. What other objections, then, can be put to me? That perhaps I am asleep (an objection I myself raised above), or that all the thoughts I have now are no more true than the dreams we have in sleep? But even if I were asleep, everything which is presented to my mind clearly, is absolutely true. And thus I recognize very clearly that the certainty and truth of all knowledge depends on the sole knowledge of the true God, so that, before I knew him, I could not know any other thing perfectly. And now that I know him, I have the means of acquiring a perfect knowledge of an infinity of things, not only relative to him, but also concerning physical nature, in so far as it serves as the object of the proofs of mathematicians, who are not concerned with its actual existence.

SIXTH MEDITATION

*Of the Existence of Material Things, and of the Real
Distinction between the Soul and the Body of Man*

THERE now remains only for me to inquire whether
material things exist, and, indeed, at least I know already
that they may exist in so far as they are considered as the
object of geometrical proofs, seeing that in this way I
perceive them very clearly and distinctly. For there is no
doubt that God has the power to produce all the things I
am capable of conceiving distinctly, and I have never
judged that it was impossible for him to do anything,
except when I found contradiction in my attempt to
conceive it correctly. Moreover, the faculty of imagination which is in me, and which I see, by experience, that
I make use of when I apply myself to the consideration of
material things, is capable of persuading me of their
existence: for when I consider attentively what imagination is, I find that it is nothing other than a certain application of the faculty of knowing to the body which is
immediately present to it, and which consequently exists.

And to make this quite clear, I note first the difference
between imagination and pure intellection or conception.
For example, when I imagine a triangle, I not only conceive it as a figure composed of three lines, but moreover
consider these three lines as being present by the power
and internal application of my mind, and that is properly
what I call imagining. Now if I wish to think of a chiliagon, I indeed rightly conceive that it is a figure composed
of a thousand sides, as easily as I conceive that a triangle is
a figure composed of only three sides; but I cannot
imagine the thousand sides of a chiliagon, as I do the three
of a triangle, neither, so to speak, can I look upon them as

present with the eyes of my mind. And although, in accordance with my custom of always making use of my imagination when I think of corporeal things, it may come about that, in conceiving a chiliagon, I picture confusedly to myself some figure, yet it is very evident that this figure is not a chiliagon, since it differs in no way from the figure which I would picture if I thought of a myriogon or of any other many-sided figure, and since it in no way serves for the discovery of the properties which constitute the difference between a chiliagon and other polygons.

But if it is a question of considering a pentagon, it is indeed true that I can conceive its figure, as well as that of a chiliagon, without the help of imagination; but I can also imagine it by applying the attention of my mind to each of its five sides, and at the same time to the space which they enclose. Thus I know clearly that I need a particular effort of the mind in order to imagine, which I do not need in order to conceive; and this particular effort of mind shows clearly the difference between imagination and pure intellection or conception.

I notice further that this power of imagination which is in me, in so far as it differs from the power of conceiving, is in no way necessary to my nature or essence, that is to say, to the essence of my mind; for, even if I did not have it, without doubt I should still remain the same as I am now, whence it seems that one can conclude that it depends on something different from the mind. And I easily conceive that if some body exists, to which my mind is joined and united in such a way that the mind can apply itself to consider it when it pleases, it may be that, by this means, it imagines corporeal objects; so that this way of thinking differs from pure intellection only in that the mind, in conceiving, turns as it were towards itself and considers some one of the ideas it has within itself; but in imagining, it turns towards the body and considers in it some thing which conforms to the idea it has formed on its own or which it has received from the senses. I easily conceive, I say, that the imagination may be formed in this way, if it is true that there are bodies; and because I can discover no other way of explaining how it is formed, I thence conjecture that it is probable that bodies exist; but this is only a probability, and although I carefully examine all things, nevertheless I do not consider that, from this distinct idea of corporeal nature which I have in my imagination, I can extract any argument which necessarily proves the existence of any body.

But I am accustomed to imagine many other things besides this corporeal nature which is the object of geometry, such as colours, sounds, tastes, pain and so on, although less distinctly. And inasmuch as I perceive these things much better through the senses, by means of which, and of memory, they seem to have reached the imagination, I believe that, in order to examine them more conveniently, it is appropriate that I should examine at the same time what sense perception is, and see if, from the ideas I receive by this mode of thinking which I call sensing, I may extract any certain proof of the existence of corporeal things.

And first I will recall to mind the things I have hitherto held to be true, as having received them through the senses, and on what grounds my belief was founded. And after that I will examine the reasons which have since obliged me to doubt them. Finally, I shall consider which of them I should now believe.

Firstly then, I perceived that I had a head, hands, feet and all the other members of which this body that I considered as a part, or perhaps also as the whole of me, is composed. Further, I perceived that this body was placed among many others, from which it was capable of receiving various agreeable and disagreeable effects, and the agreeable ones I observed by a certain feeling of pleasure, and the disagreeable ones by a feeling of pain. And besides this pleasure and pain, I also felt within me hunger, thirst and other similar appetites, as also certain composed inclinations towards joy, sadness, anger and other similar passions. And, outside myself, besides the extension, figure and movements of bodies, I noticed in them hardness, heat and all the other qualities which come under the sense of touch. Moreover, I observed in them light, colours, smells, tastes and sounds, the variety of which gave me the means of distinguishing the sky, the earth, the sea and, generally, all the other bodies, from one another.

And indeed, considering the ideas of all these qualities which were presented to my mind, and which alone I properly and immediately perceived, it was not without reason that I thought I perceived objects quite different from my thought, namely, bodies from which these ideas came. For I was conscious that these ideas were presented to my mind, without my consent being required, so that I could perceive no object, however much I wished to, if it were not present to the organ of one of my senses; and it was in no way in my power not to perceive it, when it was so present.

And because the ideas I received through the senses were much more vivid, more express and even, in their own way, more distinct than any of those which I could form for myself by meditation, or which I found imprinted in my memory, it seemed that they could not proceed from my mind, but that they had, of necessity, been caused in me by some other objects. Having no knowledge of these objects, except what the ideas themselves gave me, I could form no other conclusion than that these objects resembled the ideas which they caused.

And because I remembered also that I had used the senses rather than reason, and recognizing that the ideas which I formed of myself were not so express as those which I received through the senses, and that they were even, more often than not, made up of parts of the latter, I easily persuaded myself that I had no idea in my mind which had not passed beforehand through my senses.

Nor was it without some reason that I believed that that body which, by a special right, I call mine, belonged to me more properly and more closely than any other; for, in truth, I could never be separated from it as from other bodies; I felt in it, and by it, all my appetites and affections, and, finally, I was affected by the feelings of pleasure and pain in its parts, and not in the parts of other bodies which are separated from it.

But when I examined why from some or other feeling of pain there follows sadness of mind, and from a feeling of pleasure joy arises, or why this indefinable sensation of the stomach, which I call hunger, makes us want to eat, and dryness of the throat to drink, and so on, I could give no explanation, unless it were that nature so taught me; for there is certainly no affinity or relationship, at least that I can understand, between this sensation of the stomach and the desire to eat, any more than between the feeling of the thing which causes pain and the sense of sadness to which this feeling gives rise. And in the same way it seemed to me that I had learnt from nature all the other judgements I had formed concerning the objects of my senses, because I observed that the judgements I was in the habit of making about these objects were formed in me before I had the leisure to weigh and consider the reasons which might oblige me to make them.

But afterwards various experiences have gradually ruined all the faith I had attached to my senses. For I have observed many times that towers, which from a distance seemed round, appeared at close quarters to be square, and that huge figures erected on the summits of these towers, looked like small statues when viewed from below; and thus, in an infinity of other instances I found error in judgements based on the external senses. And not only on the external senses, but even on the internal: for is there anything more intimate or internal than pain? And yet I have sometimes heard people say, who have lost arms or legs, that they still sometimes seemed to feel pain in the limb which had been amputated; and this caused me to think that I too could not be quite certain that any one of my limbs was really affected, although I felt pain in it.

And to these reasons for doubting I have also recently added two others of a very general nature. The first is that I never believed that I could perceive anything when being awake, which I could not sometimes believe I also perceived when asleep; and as I do not believe that the things I seem to perceive while sleeping arise from objects outside myself, I did not see why I should hold this belief about those things which I seem to perceive when I am awake. And the second is that, not yet knowing, or rather feigning not to know, the author of my being, I saw nothing which could prevent me from having been so made by nature, as to be deceived even in matters which appeared to me the most true.

And as for the reasons which had hitherto persuaded me of the existence of objects of the senses, I had little difficulty in answering them. For as nature seemed to carry me towards many things from which reason turned me away, I did not believe that I ought to trust myself too much to its teachings. And although the ideas I receive through the senses do not depend on my will, I did not think that one should conclude on that account that they came from things different from myself, since perhaps some faculty might be found in me, although it is hitherto unknown to me, which caused and produced them.

But now that I am beginning to know myself better and to discover more clearly the author of my being, I do not think in truth that I ought rashly to accept all the things which the senses seem to teach us, but also I do not think that I ought to doubt them all in general.

And firstly, because I know that all the things I conceive clearly and distinctly can be produced by God precisely as I conceive them, it is sufficient for me to be able to conceive clearly and distinctly one thing without another, to be certain that the one is distinct or different from the other, because they can be placed in existence separately, at least by the omnipotence of God; and it does not matter by what power this separation is made, for me to be obliged to judge them to be different. And therefore, from the mere fact that I know with certainty that I exist, and that I do not observe that any other thing belongs necessarily to my nature or essence except that I am a thinking thing, I rightly conclude that my essence consists in this alone, that I am a thinking thing, or a substance whose whole essence or nature consists in thinking. And although perhaps (or rather as I shall shortly say, certainly,) I have a body to which I am very closely united, nevertheless, because, on the one hand, I have a clear and distinct idea of myself in so far as I am only a thinking and un-extended thing, and because, on the other hand I have a distinct idea of the body in so far as it is only an extended thing but which does not think, it is certain that I, that is to say my mind, by which I am what I am, is entirely and truly distinct from my body, and may exist without it.

Moreover, I find in me faculties of thought altogether

special and distinct from myself, such as the faculties of imagination and perceiving, without which I can indeed conceive myself clearly and distinctly as whole and entire, but I cannot conceive them without me, that is to say, without an intelligent substance to which they are attached. For in the notion we have of these faculties, or, to use the terminology of the School, in their formal concept, they comprise some sort of intellection; whence I perceive that they are distinct from me, as figures' movements and other modes or accidents of bodies are from the bodies which sustain them.

I recognize also in myself certain other faculties, such as those of changing place, of adopting various postures, and others such which cannot be conceived, any more than those mentioned above, without some substance to which they are attached, and consequently without which they cannot exist. But it is very evident that these faculties, if it be true that they exist, must be attached to some corporeal or extended substance, and not to an intelligent substance, since in their clear and distinct concept there is indeed contained some sort of extension, but no intellection at all. Further, there is in me a certain passive faculty of perception, that is to say of receiving and knowing the ideas of sensible things; but it would be useless to me, if there were not in me, or in another, an active faculty capable of forming and producing those ideas. Now this active faculty cannot be in me inasmuch as I am only a thinking thing, seeing that it does not presuppose my thinking, and also that these ideas are often presented to my mind without my contributing to it in any way, and indeed frequently against my will. This faculty must therefore necessarily be in some substance different from me, in which all the reality which is objectively in the ideas produced by it, is contained formally or eminently, as I observed above; and this substance is either a body, that is to say, a corporeal nature, in which is contained formally and in effect everything that is objectively and by representation in those ideas; or else it is God himself, or some other creature more noble than body, in which body itself is contained eminently.

But, God being no deceiver, it is manifest that he does not of himself send me these ideas directly, or by the agency of any creature in which their reality is not formally, but only eminently, contained. For, as he has not given me any faculty by which I can know that this is so, but on the contrary a very strong inclination to believe that they are sent to me or derive from corporeal things, I do not see how he could be excused of deception if in truth these ideas came from or were produced by causes other than corporeal things. And accordingly one must confess that corporeal things exist.

However, they are perhaps not exactly as we perceive them through the senses, for perception by the senses is very obscure and confused in many ways; but at least I must admit that all that I conceive clearly and distinctly, that is to say, generally speaking, all that is comprised in the object of speculative geometry, is truly to be found in corporeal things. But as concerns other things, which are either only particular, as, for example, that the sun is of such a size and shape, etc., or are perceived less clearly and distinctly, as in the case of light, sound, pain, and so on, although they are very doubtful and uncertain, nevertheless, from the fact alone that God is not a deceiver, and that consequently he has permitted no falsity in my opinions which he has not also given me some faculty capable of correcting, I believe I may conclude with assurance that I have within me the means of knowing these things with certainty.

And firstly there is no doubt that everything nature teaches me contains some truth. For by nature, considered in general, I now mean nothing other than God himself, or the order and disposition that God has established in created things. And by my nature, in particular, I understand nothing other than the composition or assemblage of all the things which God has given me.

But there is nothing that this nature teaches me more expressly, or more sensibly than that I have a body, which is ill disposed when I feel pain, which needs to eat and drink when I have feelings of hunger or thirst, etc. And therefore I must in no way doubt that there is some truth in all this.

Nature also teaches me by these feelings of pain, hunger, thirst, etc., that I am not only lodged in my body, like a pilot in his ship, but, besides, that I am joined to it very closely and indeed so compounded and intermingled with my body, that I form, as it were, a single whole with it. For, if this were not so, when my body is hurt, I would not on that account feel pain, I who am only a thinking thing, but I should perceive the wound by my understanding alone, just as a pilot sees with his eyes if any damage occurs to his ship; and when my body needs to drink or eat, I would know this simply without being warned of it by the confused feelings of hunger and thirst. For in truth all these feelings of hunger, thirst, pain, etc., are nothing other than certain confused ways of thinking, which arise from and depend on the union and, as it were, the mingling of the mind and the body.

Besides this, nature teaches me that many other bodies exist around mine, some of which I must follow and others shun. And indeed, from the fact that I perceive different sorts of colours, smells, tastes, sounds, heat, hardness, etc., I rightly conclude that there are in the bodies from which all these diverse perceptions of the senses come, certain varieties corresponding to them, although perhaps these varieties are not in fact like them.

And also, since, among these diverse perceptions of the senses, some are agreeable and others disagreeable, I can draw the certain conclusion that my body, or rather my entire self, in so far as I am composed of body and mind, can receive various pleasant or unpleasant contacts from surrounding bodies.

But there are many other things that nature seems to have taught me, which nevertheless I have not truly received from it, but which have been introduced into my mind by a certain habit I have of making judgements without proper consideration; and thus it can easily happen that they contain some error. As, for example, the opinion I have that all space in which nothing moves or makes an impression on my senses is empty; that in a hot body there is something similar to the idea of heat which is in me; that in a black or a white body there is the same blackness or whiteness which I perceive; that in a bitter or a sweet body there is the same taste or the same savour, and so on with others; that the stars, towers and all other distant bodies are of the same shape and size as they appear at a distance to our eyes, etc.

But in order that there be nothing in this that I do not conceive distinctly, I must define precisely what I properly understand when I say that nature teaches me something. For I here take nature in a more restricted meaning than when I call it the composition or assemblage of all the things God has given me, seeing that this composition or assemblage comprehends many things that belong to the mind alone, to which I do not intend to refer here in speaking of nature, as, for example, the notion I have of the truth, that what has once been done cannot any longer not have been done, and an infinity of other similar truths which I know by the natural light without the help of the body, and seeing that it comprehends also many others which belong to body alone, and are not contained here either under the name of nature, as the quality of heaviness and many others similar, of which I also do not speak, being concerned only with the things which God has given me as a being composed of mind and body. But nature, in this sense, teaches me to shun the things which cause in me the feeling of pain, and to pursue those which communicate to me some feeling of pleasure; but I do not see that, beyond this, it teaches me ever to conclude from these diverse perceptions of the senses, anything concerning external things, without the mind having carefully and maturely examined them. For it is, it seems to me, the function of the mind alone, and not of the composition of mind and body, to know the truth of these things.

Thus, although a star makes no more impression on my eye than the flame of a small torch, there is nevertheless in me no real or natural inclination that makes me believe that the star is not larger than the flame, but the fact is that I have so judged from my earliest years without any rational foundation. And although on approaching the fire I feel heat, and even pain, on approaching it a little too closely, there is however no reason which can persuade me that there is in the fire something similar to this heat, or to this pain; all I have reason to believe is that there is something in the fire, whatever it may be, which excites in me these feelings of heat or pain.

So also, although there are spaces in which I find nothing which excites or affects my senses, I must not on that account conclude that these spaces contain no body in them; for I see that, in this as in many other similar matters, I am accustomed to pervert and confound the order of nature, because although these feelings or perceptions of the senses have been placed in me only to signify to my mind which things are suitable and which damaging to the composite whole of which it is a part, and to that end being clear and distinct enough, I nevertheless use them as though they were infallible rules by which to know immediately the essence and nature of bodies which are outside me, of which they can teach me nothing that is not most obscure and confused.

But I have already sufficiently examined above how, notwithstanding the sovereign goodness of God, it happens that there is falsity in my judgements. Only one difficulty still remains concerning the things which nature teaches me to shun or avoid, and also concerning the internal senses it has put in me, for I seem to have sometimes noticed error in them and thus to be directly deceived by nature. Thus, for example, the pleasant taste of some food in which poison has been mingled may invite me to take this poison and thus to be deceived. It is true, however, that nature may be excused in this, for it leads me only to desire the food in which is an agreeable taste, and not the poison which is unknown to it; so that I can conclude nothing from this beyond that my nature does not know everything entirely and universally, concerning which there is indeed no cause for astonishment, since man, being of a finite nature, can have only knowledge whose perfection is limited. But we are also mistaken often enough, even in the things to which we are directly led by nature, as happens to sick people when they want to drink or eat what may harm them. It may be argued here, perhaps, that the cause of their error is that their nature is corrupted, but this does not remove the difficulty, because a sick man is no less truly the creature of God than a man in full health; and therefore it is as contradictory to the goodness of God that the one should have a deceitful and faulty nature as that the other should have. And as a clock, made up of wheels and counterweights, observes all the laws of nature no less exactly when it is badly made and does not keep good time, as when it completely satisfies the desire of the maker, so in the same way, if I consider man's body as being a machine, so built and composed of bones, nerves, muscles, veins, blood and skin, that although it had no mind in it, it would still move in all the same ways that it does at present, when it does not move by the direction of its will, or, consequently, with the help of the mind, but only by the disposition of its organs, I easily discern that it would be as natural for this body, supposing it had dropsy, for example, to suffer the dryness of the throat which customarily indicates to the mind the feeling of thirst, and to be disposed by this dryness to move its nerves and other organs in the way necessary for drinking, and thus to increase its sickness and to do itself harm, as it is natural, when it has no indisposition, to be prompted to drink for its good by a similar dryness of the throat. And although, having regard to the use to which the clock is destined by its maker, I may say that it is turned away from its proper nature when it does not mark the hours correctly, and, in the same way, considering the machine of the human body as having been formed by God to have in it all the movements which it is customary for the body to have, although I have reason to think that it does not follow the order of its nature when its throat is dry and drink is detrimental to its preservation, I recognize, however, that this latter sense of the term nature is very different from the other. For this is nothing more than a simple denomination depending entirely on my thought which has established a comparison between a sick man and a badly made clock, with the idea I have of a healthy man and a well-made clock, and signifying nothing which is to be found in the thing to which it is applied; whereas by the other sense of the term nature, I understand something which is truly to be found in things, and therefore which is not without some truth.

But indeed, although in respect of the dropsical body, it is only an exterior denomination when we say that its nature is corrupted, in that, without needing to drink, it nevertheless has a dry and parched throat; yet, as regards the composite whole, that is to say, the mind or the soul united with the body, it is not a pure denomination, but indeed a veritable error of nature, in that it is thirsty when it is most damaging for it to drink; and therefore there still remains to examine how the goodness of God does not prevent the nature of man, taken in this way, from being faulty and deceptive.

To begin this examination, then, I here remark firstly, that there is a great difference between mind and body, in that body, by its nature, is always divisible and that mind is entirely indivisible. For in truth, when I consider my mind, that is to say myself in so far as I am only a thinking thing, I can distinguish no parts, but conceive myself as one single and complete thing. And although the whole mind seems to be united to the whole body, yet, if a foot, or an arm, or any other part, is separated from my body, it is certain that, on that account, nothing has been taken away from my mind, nor can the faculties of wishing, feeling, perceiving, etc., properly be called its parts, for it is the same mind that is occupied, whole and entire, in willing, perceiving, conceiving, etc. But it is quite the opposite in corporeal or extended things, for there is none that I cannot easily take to pieces in thought, none that my mind cannot divide very easily into several parts, and, consequently none that I do not know to be divisible. This would suffice to teach me that the mind or soul of man is entirely different from the body, if I were not already convinced of it on other grounds.

I observe too that the mind does not receive immediately the impression from all the parts of the body, but only from the brain, or perhaps even from one of its smallest parts, namely, from the part in which this faculty called common sense is exercised, which, every time it is affected in the same way, makes the same perception arise in the mind, although meanwhile the other parts of the body may be affected diversely, as countless experiments, which it is not necessary to recall here, bear witness.

I observe, further, that the nature of the body is such that none of its parts can be moved by another part a little distant from it, which cannot also be moved in the same way by each of the intervening parts, even though the more distant part does not itself act. As, for example, in the cord *ABCD* which is tightly stretched, if the last part *D* is pulled and moved, the first part *A* will not be moved in a different way than it would be if one of the intervening parts, *B* or *C*, were pulled, and the last part *D* meanwhile remained still. And in the same way, when I feel pain in my foot, physics teaches me that this feeling is communicated by means of the nerves dispersed in the foot, which, being stretched like cords from there to the brain, when they are pulled in the foot, at the same time pull also the place in the brain whence they originate and end, and cause there a certain movement instituted by nature to make the mind aware of pain, as if this pain were in the foot. But as these nerves must pass through the leg, the thigh, the loins, the back and neck in order to stretch from the foot to the brain, it may well happen that, although their extremities in the foot are not affected, but only some of their parts which pass through the loins or the neck, this nevertheless arouses the same movements in the brain that would have been aroused there by an injury received in the foot, and hence the mind will necessarily experience the same pain in the foot as if it had received an injury there. And the same judgement must be made concerning all the other perceptions of our senses.

Finally I observe that, since each of all the movements which are in the part of the brain by which the mind is immediately affected causes merely one particular feeling, nothing better could be wished for or imagined than that among all the feelings it is capable of causing, this movement makes the mind feel the one most proper and most generally useful for the preservation of the human body, when it is in full health. But experience shows that all the perceptions which nature has given us are such as I have just said, and therefore there is nothing to be found in them which does not make apparent the power and goodness of God who produced them.

Thus, for example, when the nerves in the foot are strongly and more than usually moved, their movement, passing through the spinal cord as far as the brain, makes an impression on the mind which causes it to feel something, viz. pain, as being in the foot, by which the mind is warned and aroused to do its utmost to be rid of the cause of this pain, as being dangerous and harmful to the foot.

It is true that God could have constituted the nature of man in such a way that this same movement in the brain would have made the mind experience something altogether different: for example, that it be aware of itself, either in so far as it is in the brain, or in the foot, or in some other place between the foot and the brain, or finally that it be aware of something altogether different, whatever it might be; but nothing of all this would have contributed so well to the preservation of the body as that which the mind actually feels.

In the same way, when we need to drink, there is generated from this need a certain dryness of the throat that moves the nerves, and by means of them, the inner part of the brain, and this movement makes the mind experience the feeling of thirst, because on that occasion there is nothing more useful to us than to know that we need to drink for the preservation of our health; and likewise with other instances.

Whence it is quite manifest that, notwithstanding the sovereign goodness of God, the nature of man, in so far as it is composed of mind and body, cannot sometimes be other than faulty and deceptive.

For if there is any cause which arouses, not in the foot, but in one of the parts of the nerve which stretches from the foot up to the brain, or even into the brain itself, the same movement which ordinarily occurs when the foot is injured, pain will be felt as though it were in the foot, and the sense will naturally be deceived; for as the same movement in the brain can but cause the same feeling in the mind, and as this feeling is much more frequently aroused by a cause which hurts the foot than by one operating elsewhere, it is very much more reasonable that it should lead the mind to feel pain in the foot rather than in any other part. And although dryness of the throat does not always arise, as it usually does, from drink being necessary for the health of the body, but sometimes from quite an opposite reason, as is experienced by those with dropsy, yet it is much better that it should deceive in this case, than that, on the contrary, it should always deceive when the body is well, and the same holds true in other cases.

And indeed this consideration is of the greatest service to me not only for recognizing all the errors to which my nature is subject, but also for avoiding them, or correcting them more easily: for, knowing that all my senses more usually indicate to me what is true rather than what is false in matters relating to what is advantageous or disadvantageous to the body, and being almost always able to make use of several of these senses in the examination of the same object, and further, being able to make use of my memory to link and join present knowledge to past, and of my understanding which has already discovered all the causes of my errors, I must no longer fear henceforward that falsity may be met with in what is most commonly presented to me by my senses. And I must reject all the doubts of these last few days, as hyperbolical and ridiculous, particularly the general uncertainty about sleep, which I could not distinguish from the wakeful state: for now I see a very notable difference between the two states in that our memory can never connect our dreams with one another and with the general course of our lives, as it is in the habit of connecting the things which happen to us when we are awake. And, in truth, if someone, when I am awake, appeared to me all of a sudden and as suddenly disappeared as do the images I see when I am asleep, so that I could see neither where he came from nor where he went, it would not be without reason that I deemed him to be a spectre or phantom formed in my brain, and similar to those which are formed there when I am asleep, rather than a real man. But when I perceive things of which I clearly know both the place they come from and that in which they are, and the time at which they appear to me, and when, without any interruption, I can link the perception I have of them with the whole of the rest of my life, I am fully assured that it is not in sleep that I am perceiving them but while I am awake. And I must not in any way doubt the truth of these things, if, after having called upon all my senses, my memory and my understanding to examine them, nothing is reported to me by any one of these faculties which conflicts with what is reported to me by the others. For as God is no deceiver, it follows necessarily that I am not deceived in this.

But because the necessities of action often oblige us to

make a decision before we have had the leisure to examine things so carefully, it must be admitted that the life of man is very often subject to error in particular cases; and we must, in conclusion, recognize the infirmity and weakness of our nature.

LETTER FROM THE AUTHOR
TO THE TRANSLATOR OF THE
PRINCIPLES OF PHILOSOPHY,

to serve as a Preface

SIR,

The version of my principles that you have taken the trouble to make is so elegant and so accomplished that I expect that the work will be read by more people in French than in Latin, and will be better understood. I fear only that the title may deter some who have not been brought up to letters, or who have a poor opinion of philosophy, because the kind they were taught has not satisfied them; and this makes me think that it would be useful to add a preface, showing them what the subject of the book is, what purpose I had in writing it, and what profit may be derived from it. But although I would be the obvious person to write such a preface, since I must know these things better than anyone else, I cannot bring myself to do more than to set down in abridged form the principal points which, it seems to me, should be dealt with in a preface, and I leave it to your discretion to give to the public such part of my summary as you think fit.

I should have liked, first of all, to explain in it what philosophy is, by beginning with the most common matters, as, for example, that the word philosophy means the study of wisdom, and that by wisdom is meant not only prudence in the conduct of affairs, but a perfect knowledge of all that man can know, no less for the conduct of his life than for the preservation of his health and the discovery of all the arts, and that, in order that knowledge be of this kind, it must necessarily be deduced from first causes; so that in studying to acquire it (which is properly called philosophizing) we must begin with the search for these first causes, or principles. Now these principles must satisfy two conditions: the first, that they should be so clear and evident that the human mind cannot doubt their truth when it is applied attentively to consideration of them; the second, that it be on these principles that knowledge of other things depends, in such a way that while the principles may be known without the knowledge of other things, the converse does not hold true. Next, we must try so to deduce from these principles knowledge of the things which depend on them, that there be nothing in the whole chain of deductions deriving from them that is not perfectly manifest. Truly God alone is perfectly wise, that is to say, has complete knowledge of the truth of all things; but we may say that men have more or less wisdom as they have more or less knowledge of the most important truths. And I believe that there is nothing in all this with which all the learned do not agree.

I should then have proposed for consideration the usefulness of philosophy, and have shown that, since it extends to all that the human mind can know, we must believe that it alone distinguishes us from savages and barbarians and that each nation is the more civilized and polished the better its members are versed in philosophy and, accordingly, that the greatest good which can exist in a State is to have true philosophers. Besides this, I should have shown that, as far as each individual is concerned, it is not only useful for him to live among those who apply themselves to this study, but that it is incomparably better to apply himself to it; just, no doubt, as it is much better to use one's own eyes to guide oneself, and to enjoy by the same means the beauty of colour and light, than to have them closed and to follow the lead of another, although this latter course is better than to keep them closed and to have only oneself as a guide. For to live without philosophizing is indeed tantamount to keeping one's eyes closed without ever trying to open them, while the pleasure of seeing all the things that our sight reveals is not to be compared to the satisfaction which comes from the knowledge of those discovered by philosophy. And, finally, this study is more necessary for regulating our manners, and for guiding us in this life, than the use of our eyes is for directing our steps. Brute beasts, which have only their bodies to preserve, are occupied continuously in seeking food; but men, whose principal part is the mind, should make the search for wisdom their main preoccupation, for wisdom is the true nourishment of the mind; and I am convinced, moreover, that there are many who would not fail to do so, if only they had hope of succeeding, and knew how far they are capable of it. There is no soul, provided it has the least spark of nobility, which remains so strongly attached to the objects of the senses, that it does not sometimes turn away from them to aspire to some greater good, even though it frequently does not know in what that good consists. Those whom fortune favours most, who have an abundance of health, of honours, and of riches, are no more exempt from this desire than the others; on the contrary, I am persuaded that it is they who yearn the most ardently after another good, more sovereign than all those which they possess. But this sovereign good, considered by natural reason without the light of faith, is nothing other than the knowledge of truth through its first causes, that is to say wisdom, of which philosophy is the study. And as all these things are wholly true, it would not be difficult for them to convince if they were properly deduced.

But, as one is prevented from believing them by experience, which shows that those who profess to being philosophers are often less wise and reasonable than others who have never applied themselves to this study, I should have explained here summarily in what our present knowledge consists and what are the degrees of wisdom which we have reached. The first degree contains only notions so clear of themselves that they can be acquired without meditation; the second comprises all the knowledge acquired through the experience of the senses; the third, what we learn from intercourse with others; to which may be added, as the fourth, reading, not of all books, but particularly of those which have been written by people capable of giving us good instruction, for it is a kind of conversation that we have with their authors. And it seems to me that all the wisdom we commonly have is acquired only by these four means; for I do not put forward divine revelation here, because it does not lead us by degrees, but raises us at a stroke to an infallible belief. Now there have been at all times great men who have tried to find a fifth degree by which to arrive at wisdom, incomparably higher and more assured than the other four: this way is to seek the first cause and true principles from which can be deduced the reasons of all that can be known to man; and it is particularly those who have strived to do this who are called philosophers. However, I am not aware that there has been up to now any one of them who has succeeded in this enterprise. The first and principal of those whose writings we possess, are Plato and Aristotle, between whom there was no difference, except that the former, following in the steps of his master, Socrates, ingenuously confessed that he had not yet been able to find anything certain, and contented himself with writing what seemed to him to be probable, imagining, to this end, certain principles by which he tried to account for the other things. Aristotle, on the other hand, was less frank and although he had been Plato's disciple for twenty years, and had no other principles than his, he entirely changed his way of stating them, and proposed them as true and certain, although it is most unlikely that he ever thought them to be so. But these two men were very intelligent and had a great deal of the wisdom which is acquired by the four means described above; and this gave them so much authority that those who came after them laid more store on following their opinions than on seeking something better. The principal point of debate among their disciples was as to whether all things should be doubted or whether there were some which were certain. And this dispute led them on both sides into extravagant errors; for some of those who were for doubt, extended it even as far as the actions of everyday life, thus neglecting to use prudence in their conduct; and those who maintained the existence of certainty, on the supposition that they could depend on their senses, placed all their trust in them, to the extent that it is said of Epicurus that he dared to affirm, against all the reasonings of the astronomers, that the sun is no larger than it appears. It is a fault which can be observed in most disputes, that, truth being mid-way between the two opinions that are held, each side departs the further from it the greater his passion for contradiction. But the error of those who leaned too much to the side of doubt was not followed for long, and the error of the others has been corrected somewhat in that recognition has been given to the fact that the senses deceive us in many things. All the same, I am not aware that this error has been entirely removed by showing that certainty is not in the senses, but in the understanding alone, when it has evident perceptions; and that, while we have only the knowledge which is acquired by the first four degrees of wisdom, we must not doubt the things which appear to be true in what concerns the conduct of life, nor must we consider them to be so certain that we cannot change our opinion when obliged to do so by the light of some reason. For lack of knowing this truth, or – if there were some who knew it – for lack of making use of it, most of those who, in these last centuries, have aspired to be philosophers, have blindly followed Aristotle, with the result that they have often corrupted the sense of his writings, by attributing to him various opinions which he would not recognize as being his if he were to come back to this world; and those who have not followed him (among whom have been many of the best minds) did not yet fail to be imbued with his opinions in their youth, since these are the only opinions taught in the schools, and these opinions so prejudiced their minds that they were unable to arrive at the knowledge of true principles. And although I esteem all philosophers and do not wish to make myself odious by criticizing them, I can give a proof of my assertion which I do not believe any one of them will reject, namely, that they have all put forward as a principle something they have not known perfectly. For example, I know none of them who did not suppose heaviness to be a property of terrestrial bodies; but, although experience shows us very clearly that bodies we call heavy fall towards the centre of the earth, we do not on that account know what is the nature of this thing we call heaviness, that is to say, of the cause or principle which makes them fall in this way, and we must learn about it from another source. The same can be said of the void, of atoms, of heat and cold, of dryness and humidity, and of salt, sulphur and mercury, and all the other similar things that some have posited for their principles. But the conclusions deduced from a principle that is not evident cannot themselves be evident, although they may be deduced from it evidently; from which it follows that none of the reasonings based on such principles could give them certain knowledge of anything, nor consequently advance them one step in the search for wisdom. And if they have found some truth or other, it was only by some of the four means set out above. Nevertheless, I have no desire to diminish the honour which each of them can claim; I am only obliged to say, for the consolation of those who have not studied, that just as in travelling, when we turn our backs on the place to which we wish to go, we go the further away from it the longer and the quicker we walk, with the result that, although we may afterwards be put on the right path, we cannot arrive as quickly as if we had not walked at all, so, when we have false principles, the more we cultivate them and the greater the care with which we apply ourselves to drawing various conclusions from them, thinking that we are philosophizing well, the further we are departing from knowledge of truth and wisdom. Hence we must conclude that those who have learned the least of all that has hitherto been called philosophy are the most capable of learning the true one.

After making these things clear, I should have liked to set down here the reasons which serve to prove that the principles by which we may arrive at this highest degree of wisdom, in which consists the sovereign good of human life, are those which I have put in this book; and two reasons alone are sufficient to establish this, the first of which is that they are very clear, and the second, that all other truths can be deduced from them; for these two conditions only are required of true principles. Now I easily prove that they are very clear; firstly, by the way in which I have found them, namely by rejecting everything in which I encountered the least occasion for doubting, for it is certain that those which could not be rejected in this way are, when attentively considered, the most evident and clear that the human mind can know. Thus, by considering that he who decides to doubt everything cannot nevertheless doubt that he exists while he doubts, and that what reasons thus, in not being able to doubt itself and doubting nevertheless all the rest, is not what we call our body, but what we call our soul or thought, I have taken the being or the existence of this thought for the first principle, from which I very clearly deduced the following truths, namely, that there is a God who is the author of all that is in the world, and who, being the source of all truth, has not created our understanding of such a nature as to be deceived in the judgements it forms of the things of which it has a very clear and distinct perception. Those are all the principles of which I make use concerning immaterial or metaphysical things, from which I deduce very clearly the principles of corporeal or physical things, namely, that there are bodies extended in length, breadth and depth, which have diverse shapes and move in various ways. Such are, in brief, all the principles from which I deduce the truth of all other things. The second reason which proves the clarity of these principles is that they have been known in all times, and even accepted as true and indubitable by all men, with the exception only of the existence of God, which has been doubted by some, because they attributed too much to the perceptions of the senses, and God can be neither seen nor touched. But although all the truths which I count among my principles have been known in all times and by all men, nevertheless, there has been no one until now, as far as I know, who has recognized them as the principles of philosophy, that is to say, as such that we can deduce from them knowledge of everything else in the world. This is why it remains for me here to prove that they are such; and it seems to me that I cannot do it better than through experience, that is to say, by inviting readers to read this book. For although I have not treated of all matters in it, as this would have been impossible, I think I have so explained everything of which I have had occasion to treat, that those who read it attentively will have cause to be convinced that there is no need to seek other principles than those I have given, in order to arrive at the highest knowledge of which the human mind is capable; especially if, after having read my writings, they take the trouble to consider how many diverse questions are explained in them, and if, perusing the works of others, they see of what scant probability are the reasons adduced to explain the same questions by principles different from mine. And, so that they may the more easily undertake this, I might have said that those who are imbued with my doctrines have much less trouble in understanding the writings of others and in assessing their true value, than those who are not so imbued; and this is quite the contrary of what I said earlier of those who began with the ancient philosophy, namely that the more they have studied it, the less fit they usually are for grasping the true one.

I should also have added a word of advice concerning the way to read this book, which is, that I should wish it first to be read straight through completely, like a novel, without the reader straining his attention too much or stopping at difficulties he may encounter, in order simply to know in broad outline what the matters are of which I treat; and that afterwards, if he considers them to merit further examination, and has the curiosity to know their causes, he may read it a second time in order to observe the development of my reasonings; but that he must not then give it up in despair, if he cannot follow it completely throughout or understand all the reasonings; he has only to mark with a stroke of his pen the places where he comes across difficulties and to continue to read without interruption to the end; then if he takes up the book for the third time, I feel sure that he will find the solution of most of the difficulties he marked before; and that, if any still remain, their solution will eventually be found in a further reading.

I have observed, in examining the natural constitution

of many minds, that there are almost none so gross or so slow as to be incapable of acquiring correct opinions and even of reaching the highest knowledge, if they are guided correctly. And this can also be proved by reason; for, as the principles are clear and as nothing should be deduced from them except by very evident reasoning, a man always has enough intelligence to understand the conclusions which stem from them. But, besides the hindrance of prejudices, from which no one is entirely exempt, although it is those who have most studied the false sciences who are the most hindered by them, it almost always happens that people of moderate intelligence neglect to study, because they believe themselves incapable of it, and that others, more ardent, press on too quickly: whence it arises that they often accept principles which are not evident, and draw uncertain conclusions from them. This is why I would like to assure those who lack confidence in their powers, that there is nothing in my writing that they cannot fully understand, if they take the trouble to examine them; and at the same time I would also warn the others, that even the best minds will need much time and attention to observe all that I designed to embrace in these writings.

After this, in order to make clear my aim in publishing them, I should like to explain here the order which it seems to me should be followed in order to instruct oneself. Firstly, a man who as yet has only the vulgar and imperfect knowledge that can be acquired by the four means explained above, must before all else form for himself a moral code which may suffice to regulate the actions of his everyday life, because this brooks no delay, and because we must above all try to live well. After that, he must also study logic: not that of the Schoolman for it is only, properly speaking, a dialectic which teaches the means of conveying to others the things we know already, or even of talking a lot without judgement about what we do not know and which, consequently, corrupts rather than increases good sense; but the logic which teaches us to conduct our reasoning aright in order to discover the truths of which we are ignorant; and, because it depends largely on practice, it is desirable that he exercise himself for a long time in practising its rules on easy and simple questions, like those of mathematics. Then, when he has acquired some facility in discovering the truth in these questions, he must begin seriously to apply himself to true philosophy, of which the first part is metaphysics, which contains the principles of knowledge, among which is the explanation of the principal attributes of God, of the immateriality of the soul, and of all the clear and simple notions that are in us. The second is physics, in which, after having discovered the true principles of material things, we examine, in general, how the whole universe is composed; then, we examine, in particular, the nature of the earth and of all the bodies which are to be found most commonly about it, as air, water, fire, the lodestone and other minerals. Following this, it is necessary also to examine in particular the nature of plants, of animals, and above all, of man, in order to be able thereafter to discover the other sciences useful to us. Thus, all philosophy is like a tree, whose roots are metaphysics, the trunk physics and the branches which grow out of this trunk are all the other sciences, which are reduced to three principal ones, namely, medicine, mechanics and ethics, by which I understand the highest and most perfect science which, as it presupposes a complete knowledge of the other sciences, is the last degree of wisdom.

But as it is not from the roots or the trunk of trees that the fruits are picked, but only from the extremities of their branches, so the principal usefulness of philosophy depends on those of its parts which we can learn only last of all. But although I am ignorant of almost all of them, the zeal I have always felt for endeavouring to render service to the public is the reason why I published, some ten or twelve years ago, certain essays on the things I thought I had learned. The first part of these essays was a *Discourse On the Method of Properly Conducting One's Reason and of Seeking the Truth in the Sciences*, in which I gave a summary of the principal rules of logic and of an imperfect ethic that may be followed provisionally so long as one does not yet know of a better. The other parts were three treatises: one on *Dioptrics*, another on *Meteors*, and the last on *Geometry*. In the *Dioptrics*, I attempted to show that one could go far enough in philosophy to arrive, by means of it, at a knowledge of the arts which are useful to life, because the invention of the telescope, which I explained in this treatise, is one of the most difficult that has ever been sought. In the *Meteors*, I attempted to make apparent the difference between the philosophy that I cultivate, and that taught in the Schools, in which the same matters are usually treated. Finally, in the *Geometry*, I aimed to demonstrate that I had discovered many things which were hitherto unknown, and thus to give ground for believing that many others may still be discovered, in order to incite all men to seek for the truth. Since that time, foreseeing the difficulty that many would have in grasping the foundation of metaphysics, I tried to explain the principal points concerning them in a book of *Meditations*, which is not very large but the size of which has been increased, and the content much clarified, by the objections which several very learned persons sent me on the subject, and by the replies which I made to them. Then, finally, when it seemed to me that these earlier treatises had sufficiently prepared the minds of my readers to accept the *Principles of Philosophy*, I published it also, dividing the book into four parts, the first of which contains the principles of knowledge, which is what may be called first philosophy or metaphysics; and that is why, in order to be able to understand it fully, it is appropriate first to read the Meditations which I wrote on the same subject. The three other parts contain all that is most general in physics, namely, the explanation of the first laws or principles of nature, and the way in which the heavens, the fixed stars, the planets, comets and generally the whole universe is composed; then, in particular, the nature of this earth, of air, water, fire and the lodestone, which are the bodies most commonly to be found everywhere about it, and of all the qualities to be observed in these bodies, such as light, heat, weight, and so on. Thus I think I have begun to explain the whole of philosophy in an orderly way, without having omitted any of the matters which should precede the last to be treated. But in order to bring this plan to its conclusion, I ought hereafter to explain in the same way the nature of each of the other more particular bodies which are on the earth, namely, minerals, plants, animals and especially man; finally, I ought to treat exactly of medicine, ethics and mechanics. This is what I should do in order to give mankind a complete body of philosophy; and I do not yet feel myself so old, I do not so distrust my powers, I do not find myself so far removed from the knowledge of what remains, that I should not dare undertake the completion of this design, had I but the means to make all the experiments I should need in order to support and justify my reasonings. But realizing that this would require great expenditure beyond the resources of a private individual like myself, unless he were aided by the public, and as I have no reason to believe that such aid would be forthcoming, I believe that I ought from now on to content myself with studying for my own instruction, and that posterity will excuse me if I fail to work further for it.

However, so that it may be seen how I think I have already served posterity, I will mention here what the fruits are that I am persuaded can be gathered from my principles. The first is the satisfaction which will be felt on finding in them many truths hitherto unknown; for, although frequently truth does not excite our imagination as much as falsities and fictions, because it seems simpler and less wonderful, yet the contentment it gives is always more lasting and solid. The second fruit is that in studying these principles one becomes accustomed little by little to judge better all the things one encounters, and thus to become wiser; and in this the principles will have the opposite effect to that of the common philosophy, for it can easily be observed in those we call pedants that the common philosophy renders them less capable of reason than they would be if they had never learnt it. The third is that the truths which they contain, being very clear and certain, will remove all causes of dispute and thus dispose men's minds to gentleness and concord; whereas, on the contrary, the controversies of the Schools which, insensibly rendering those who practise them more wrangling and obdurate, are perhaps the prime cause of the heresies and dissensions which now afflict the world. The last and chief fruit of these principles is that one will be able, in cultivating them, to discover many truths that I have not dealt with, and thus passing gradually from one to another, to acquire in time a perfect knowledge of the whole of philosophy and to rise to the highest degree of wisdom. For, just as one sees in all the arts that, although at the beginning they are crude and imperfect, yet because they contain something true and whose effect is shown by experience, they are perfected gradually through use; so in philosophy, when we have true principles, we cannot fail by following them sometimes to meet with other truths; and we could not better prove the falsity of the principles of Aristotle, than by saying that men have been unable to make any progress by their means during the many centuries that these principles have been followed.

I well know that there are some men who are over hasty and use so little circumspection in what they do, that, even with perfectly solid foundations, they could not build anything firm; and because it is usually they who are the readiest to write books, they could in a short time spoil all that I have done, and introduce into my way of philosophizing, whence I have carefully tried to banish them, uncertainty and doubt, if their writings were to be accepted as mine, or as containing my opinions. I had quite recently an example of this in one of those who were believed to want to follow me the most closely, a man of whom I had even written somewhere that I was so sure of his mind that I did not think he had any opinion that I would not be happy to acknowledge as my own; for he published last year a book entitled *Fundamenta Physicae*, in which, although he appears to have put nothing, concerning physics and medicine, that he has not taken from my writings, both from those I have published and from another still imperfect concerning the nature of animals, which fell into his hands, yet, because he has transcribed them badly, and changed the order, and denied certain truths of metaphysics, on which all physics must be based, I am obliged to disavow his work completely, and here to request my readers not to attribute to me any opinion, unless they find it expressly in my writings, and not to accept any opinion as true, either in my writings or elsewhere, unless they see it very clearly to be deduced from true principles.

I well know also that many centuries may pass before

all the truths that can be deduced from these principles have been so deduced, because most of those which remain to be discovered depend on certain particular experiments which will never be encountered by chance, but which must be sought after with care and expense by men of the highest intelligence, and because it is unlikely that the same people who have the ability to make good use of them will have the means of performing them, and also because most of the best minds have conceived so poor an opinion of all philosophy, on account of the defects they have observed in the philosophy which has been in use up to now, that they cannot apply themselves to the search for a better one. But if, finally, the difference which they will see between my principles and all those of others, and the great succession of truths that can be deduced from them, cause them to recognize how important it is to continue in the search for these truths, and to what degree of wisdom, to what perfection of life, to what happiness they can lead, I venture to believe that there will be no one who will not try to devote his labours to so profitable a study, or at least to support and be willing to help, to the extent of his power, those who devote themselves to this task with profit. My wish is that posterity may witness the happy outcome of it, etc.

FOR THE BEST IN PAPERBACKS, LOOK FOR THE 🐧

In every corner of the world, on every subject under the sun, Penguin represents quality and variety – the very best in publishing today.

For complete information about books available from Penguin – including Pelicans, Puffins, Peregrines and Penguin Classics – and how to order them, write to us at the appropriate address below. Please note that for copyright reasons the selection of books varies from country to country.

In the United Kingdom: Please write to *Dept E.P., Penguin Books Ltd, Harmondsworth, Middlesex, UB7 0DA*

If you have any difficulty in obtaining a title, please send your order with the correct money, plus ten per cent for postage and packaging, to *PO Box No 11, West Drayton, Middlesex*

In the United States: Please write to *Dept BA, Penguin, 299 Murray Hill Parkway, East Rutherford, New Jersey 07073*

In Canada: Please write to *Penguin Books Canada Ltd, 2801 John Street, Markham, Ontario L3R 1B4*

In Australia: Please write to the *Marketing Department, Penguin Books Australia Ltd, P.O. Box 257, Ringwood, Victoria 3134*

In New Zealand: Please write to the *Marketing Department, Penguin Books (NZ) Ltd, Private Bag, Takapuna, Auckland 9*

In India: Please write to *Penguin Overseas Ltd, 706 Eros Apartments, 56 Nehru Place, New Delhi, 110019*

In Holland: Please write to *Penguin Books Nederland B.V., Postbus 195, NL–1380AD Weesp, Netherlands*

In Germany: Please write to *Penguin Books Ltd, Friedrichstrasse 10–12, D–6000 Frankfurt Main 1, Federal Republic of Germany*

In Spain: Please write to *Longman Penguin España, Calle San Nicolas 15, E–28013 Madrid, Spain*

In France: Please write to *Penguin Books Ltd, 39 Rue de Montmorency, F-75003, Paris, France*

In Japan: Please write to *Longman Penguin Japan Co Ltd, Yamaguchi Building, 2–12–9 Kanda Jimbocho, Chiyoda-Ku, Tokyo 101, Japan*

PENGUIN CLASSICS

Honoré de Balzac	**Cousin Bette**
	Eugénie Grandet
	Lost Illusions
	Old Goriot
	Ursule Mirouet
Benjamin Constant	**Adolphe**
Corneille	**The Cid/Cinna/The Theatrical Illusion**
Alphonse Daudet	**Letters from My Windmill**
René Descartes	**Discourse on Method and Other Writings**
Denis Diderot	**Jacques the Fatalist**
Gustave Flaubert	**Madame Bovary**
	Sentimental Education
	Three Tales
Jean de la Fontaine	**Selected Fables**
Jean Froissart	**The Chronicles**
Théophile Gautier	**Mademoiselle de Maupin**
Edmond and Jules de Goncourt	**Germinie Lacerteux**
J.-K. Huysmans	**Against Nature**
Guy de Maupassant	**Selected Short Stories**
Molière	**The Misanthrope/The Sicilian/Tartuffe/A Doctor in Spite of Himself/The Imaginary Invalid**
Michel de Montaigne	**Essays**
Marguerite de Navarre	**The Heptameron**
Marie de France	**Lais**
Blaise Pascal	**Pensées**
Rabelais	**The Histories of Gargantua and Pantagruel**
Racine	**Iphigenia/Phaedra/Athaliah**
Arthur Rimbaud	**Collected Poems**
Jean-Jacques Rousseau	**The Confessions**
	Reveries of a Solitary Walker
Madame de Sevigné	**Selected Letters**